Bangernomics Rebooted

James Ruppert

ISBN: 978-1-9160522-1-5

DEDICATION

To my Father who always took me with him to buy cars and taught me so much and my Mother who put up with a boy who filled her house with car parts. For the Rebooted Edition this is for Olivia, who will carry on where I leave off.

PUBLISHER'S NOTE

Whilst the Publisher and Author have made every effort to ensure the accuracy of the information contained within this book, no liability can be accepted for any car purchase made. Readers should note, in particular, that the information about car types can only be given in the most general terms and that individual models can and will vary according to their earlier treatment. In the case of any doubt regarding the condition or safety of the proposed purchase, readers should seek professional advice and not proceed with any purchase where doubt remains.

CONTENTS

ACKNOWLEDGEMENTS

A number of people gave me help, or paid a big enough bribe for inclusion in this section. Thank you to: Wembley Car Auctions; Handmade Films; Nigel and Karen Biles for the loan of `Emmerson' their red Volvo; Rachel Ponter for driving her Honda Prelude in the rain; Alan Gardiner for the loan of his Morris Marina; Ted Sparrow at the East Anglian Mini Centre (047337 771) for letting me use his ramp and photograph his cars. The biggest thank you of all goes to Dee for being a part-time model and full time PA.

ORIGINAL INTRODUCTION

Bangernomics is all about motoring at the blunt end, but it wasn't my idea*. In the most extreme form Bangernomics was first practised by an acquaintance of my Father who back in the 1950s bought £5 Ford Populars, never opened the bonnet and simply added petrol. After a year or so they broke down and he left the 'Pop' where it stopped.

Since then, thousands of others have realised that running a cheap car makes a lot of common and financial sense. However, many fall at the first hurdle when they buy an unsuitable heap. Yet the principles of buying wisely are simple enough, which is where *Bangernomics* comes in. There are no excuses for acquiring an MOT borderline or deathtrap when you could buy a good value car that suits your requirements. So let *Bangernomics* be your guide. Enrolment for the Bangernomics course is the paltry cost of this book and if you stick with it, you can earn a highly coveted Bangernornics Proficiency Certificate (details at the back). Don't forget that at all times throughout this course you are in safe, if slightly shaky hands. Because when it comes to buying cars I have made every possible mistake. I hope you can learn from them.

James Ruppert July 1992

*When I said that Bangernomics wasn't my idea, I lied, because the word is. That means the inevitable sequel, Bangernomics II, T-shirt, video and motion picture rights, 'Slog the Bangernomic Dog' cartoon series, all belong to me and the publishers, so don't infringe the copyright unless you cut me in to the tune of 60% or you'll be hearing from my lawyers.

REBOOTED INTRODUCTION

It is 26 years since the first Bangernomics Book was published and 29 years since I first used the phrase in a magazine article. Since then, if I had a pound for every time someone told me the book had changed their motoring lives, I'd live in a much bigger house with a far larger garage. Also, if that many people read it, why wasn't it top of the bestseller lists? Never mind, lots of people liked the original but they can't buy it anymore. Luckily the original publishers, Ian Allan, gave me back the rights to Bangernomics when I asked nicely many years ago. So here it is, full of dated information, badly taken pictures and silly cartoons. In many ways it is a time capsule of early '90s used car buying. An era when the streets still contained Ford Cortinas, real minis and Morris Marinas. The Bangernomics bottom line is that you won't learn anything new. It might though amuse, entertain and annoy which makes going to all the bother of reformatting Bangernomics, worth it. Here is Bangernomics Rebooted.

James Ruppert 2019

1. THE BIG BANG THEORY: BANGERNOMICS EXPLAINED

Do you want to cut your motoring costs?

Do you want to know how to buy a good used car?

Answer yes to either of those questions and it is a certainty that you need a course in Bangernomics. The concept is simple enough and is sometimes called the Kleenex principle: buy a cheap car, run into the ground, throw away, and buy another. Not convinced? Well, pencil yourself into this all too familiar story. Simply insert your Christian name, or that of the person you have bought the book for (   ) , so that they/you can see where you/they might/have gone wrong. This is a true story.

Once upon a time (   ) was standing on a garage forecourt nervously peering into a showroom full of shiny new, and nearly new cars. (   ) ventures inside for a closer look and comes face to face with windscreen sticker prices that dwarf most third world debts. He/she reels from the shock and runs toward the door. But before (   ) reaches safety, a salesman jumps out from behind a potted rubber plant and traps him/her in the twilight zone of motoring dreams and illusions. What thought was beyond his/her financial grasp has been chopped up into seemingly manageable monthly payments. Minutes later he/she has signed on the dotted line and is sitting inside something, shiny and expensive. (   ) presses the button for the electric window and suddenly the noises and smells of the real world bring him/her out of the trance.

It suddenly occurs to ( ) that the electric window was a costly option and he/she rummages furiously through the paperwork. 'Did I seriously need metallic paint, an eight speaker stereo, or a coach-line?' asks him/herself and how come you have to pay delivery and preparation charges on top of the retail price? ( ) felt sick, this was not a dream, but a nightmare. He/she swallowed hard, driving jerkily and unhappily into the sunset. Something told he/she would be back.

The thing that ( ) told to go back was the bank manager. He/she'd had enough of monthly payments, the labour charges for routine servicing, the costly comprehensive insurance and never having enough money left over for food. ( ) had not used it that much and anyway, there is not much point in owning all-terrain, turbo-powered van if all you are going to do is pop down to Tescos and visit mum was battered, bruised and the bank account empty, so without any fear he/she cornered the salesman against a pile of brochures and demanded he buy the car back. Without a hint of embarrassment he offers less than half what originally paid. Apparently, everyone wants the new model now and no one can live with that heliotrope orange paint. There is also the tricky question of finance to be settled. He helpfully suggests that part exchange it for something else, which 'costs just a few pounds more a month'. ( ) started to tremble. The salesman then suggests selling the vehicle privately for a better price, but ( ) has been advertising it for weeks without any response. ( ) stumbles back outside into the daylight and bumps into a friend who says 'I need a cheap set of wheels, you know about motors, what should I do?

( ) is the perfect student for a course in Bangernomics.

Bangernomics contrasts the absurd expense of buying new, or almost new, with the supreme good sense of buying used. Once upon a time you really could buy a car for a few quid, just add petrol and when it came to a spluttering halt, abandon it by the roadside. Today however, a car must be roadworthy, thanks to the MOT, and dumping is hardly environmentally friendly. For the responsible and cost conscious motorist, Bangernomics is the answer and this book has just a few simple aims:

1. To de-mystify the art of buying a used car.

2. To save money.

3. To promote safe motoring.

4. To cure the common cold

If you still have some lingering doubts, we surveyed a cross section of car owners about Bangernomics and they came up with the following questions

Answer: In both the short and long term a Bangernomic approach to car ownership makes sound financial sense. Purchasing a new car involves immediate financial losses. Before it has turned a wheel the VAT element that make up a large proportion of the purchase price are written off. Add to that the dealer's charge to put the car on the road and you can start to feel rather unwell. But the bad news does not end there, depreciation takes an unhealthy bite out of the car's value too. In the first year where the drop is worst you can figure on depreciation being worth at least 50% of the original purchase price, often it can be even higher than that. With a banger, all the previous owners have taken the brunt of depreciation and high initial running costs. That leaves you with a car that is unlikely to drop in value, having reached rock bottom. Even if you did sell in 12 months time what would you rather have, 50% of £500 or 50% of £12,000?

Answer: Difficult one that. To some people a car becomes a banger once the ashtrays are full. Others think that it is last year's now unfashionable model. However, to the student of Bangernomics, it is a car that is attractively cheap which still has some life left in it. Ideally it will be a car that you can afford, rather than borrowing large sums in order to buy. Cars will usually be over five years old and probably closer to celebrating their tenth birthday. See the A to B chapter for further details about what constitutes the perfect banger. Bear in mind though that the tips, wrinkles and advice contained in Bangernomics applies equally to all car purchases, after all we want to sell as many copies of this book as possible.

Answer: Anyone who is car proud and spends most of Sunday with a bucket, sponge and polish. Motoring snobs who have to keep up with the Joneses, must have the latest, most expensive model and constantly remind everyone how much they paid for it.

Answer: No. Let's take servicing. Older cars are simpler to work on and therefore cheaper to maintain, especially if you carry out the basics yourself. There are also alternative sources for parts rather than paying through the nose or, as it is better-known, main agent prices. This book will show the mechanically cack-handed how to do the minimum maintenance for maximum mileage. With new cars you don't get that option. Also a Bangernomic vehicle only merits third party insurance rather than costly comprehensive cover that a new car demands. On the same subject, a dent to a banger might even improve its looks and an expensive visit to the bodyshop is unnecessary. There is also little reason to worry about the bump and grind of close quarter city traffic and no lost sleep that a vandal might add another scratch to the paintwork.

Answer: For you environmentally concerned little

Greenies out there, it is and it isn't. When you consider the huge amount of pollution that the internal combustion engine causes, car ownership of any kind has to be viewed as irresponsible. However, a Bangernomic approach means that a car can be recycled, rather than abandoned. The natural resources and energy used to make a new car is phenomenal, which makes prolonging the life and disposing responsibly of existing machinery decidedly 'green'. So put your bike in the shed and get motoring.

Answer: Of course. But then so is the purchase of a new car, which may break down within days leaving the buyer with precious few legal remedies. Even a nearly new, warranted used car may have been abused and neglected during its life leaving the buyer with an expensive liability. So what's so dodgy about buying at the bargain basement end of the market? If you sincerely want to save money Bangernomics will take some of the uncertainties out of a banger purchase and put a satisfied smile on your face.

Answer: Anyone who is car proud and spends most of Sunday with a bucket, sponge and polish. Motoring snobs who have to keep up with the Joneses, must have the latest, most expensive model and constantly remind everyone how much they paid for it.

For the completely sceptical, here is the bottom line. This table was compiled to show the differences between running an old and new car. The insurance quotes were based on a 30 year old clerk with a full no claims bonus living in Middlesex. The service charges relate to an average mileage of 12,000 miles and two visits to a garage at 6,000 and 12,000 miles. Peugeot finance charges are based on a package offered by dealers in 1992, and consisted of 10%

deposit with balance paid over four years at 7.9%. Depreciation based on prevailing trends in 1992, the Peugeot being sold privately. The Cortina figure may be pessimistic, but allows for deterioration and repairs over the year. In some cases it could even be worth more! Spares costs for Peugeot did not include labour charge for fitting.

1982 Peugeot 405 1.6 Style • Price £9899 • Delivery Charge £385 • **ANNUAL COSTS** - Road Tax £120 • Depreciation £3000 • MOT n/a • Insurance £470 • Cost of Finance £730 • Servicing £220 **TOTAL LOSS** £4540 **SPARES COSTS** - Engine (exchange) £1380 fitted • Gearbox (exchange) £580 fitted • Clutch £120 fitted • Exhaust £190 fitted • Starter Motor £124fitted • Front Wing £70.

By the time you read this, the figures will undoubtedly have changed, but that is not relevant. The point is that a Bangernomic approach saves money at every stage. Why not do your own calculations?

1982 Ford Cortina 1.6L • Price £500 • Delivery Charge n/a • **ANNUAL COSTS** - Road Tax £120 • Depreciation £150 • MOT £20 • Insurance £194 • Cost of Finance n/a (cash) • Servicing £80 **TOTAL LOSS** £564

SPARES COSTS - Engine (exchange) £240 fitted • Gearbox (exchange) £90 fitted • Clutch £60 fitted • Exhaust £38.95 fitted • Starter Motor £50 fitted • Front Wing £42, but secondhand £15-£25

If you have read this far and not bought the book. Buy it now. Then go home, get a pencil, pen, calculator, telephone and strap yourself into the armchair for a roller coaster ride into the thrills, spills, trials and tragedies of buying a banger. Good luck. Everyone is capable of completing the Bangernomics course. At the end of this book you will find details about obtaining a Bangernomics certificate and put the letters B.A.N.G.E.R. after your name. But first you have to buy the car. After the theory lets get down to some practice. PTO

2. BANGERS & CASH: BUDGETING FOR A BANGER

The true Bangernomics student tips up their piggy bank and shakes out the remaining coppers to find out what their used car budget really is. If you don't know how much you can afford it will certainly all end in tears. So you must carry out some basic budgeting, not only to arrive at a sum which goes towards the purchase, but also a weekly allowance that helps you to run the car.

PURCHASE PRICE: The logic is simple, how much do you want to spend? £ How much do you have saved? £( ) If there is a shortfall, then think again. If there is a surplus, is that enough to constitute 'savings' given your lifestyle and requirements? Now is the time to make up your mind. If you can't afford it, then forget about buying for a few months. The Bangernomics approach is to do a bit of overtime, save up and use your own hard earned cash to qualify for the course. If nothing else, it concentrates your mind wonderfully on making a purchase that you won't regret. Even if you make a mistake and buy a dud then at least you won't be paying for it over the next 12 months as you would with a loan. Once you have decided to spend a certain amount on the car write it down in this book in ink. Don't even think about changing that figure unless you win the pools. As you will discover, the big problem with buying a car is that you can get carried away and spend more than you have budgeted, thanks to the slimy advances of a slick salesman, or your own weakness when your heart overrules your head and authorises a costly impulse buy.

MY BUYING BUDGET IS £ .... (in ink please)

SHOPPING FOR FINANCE: If you cannot quite manage to save enough, or there is little likelihood of doing so, then there are some alternatives. The big problem about borrowing money is that in most cases it costs you even more money in terms of interest charges. The most important thing to do is look at the total cost of the loan, that is the amount that will be repayed. These can be very murky and confusing waters indeed.

• Firstly look at the APR (annual percentage rate), the actual interest you pay on the loan. Clearly a 20% APR is better than a 25%.

• Don't base the decision solely on the APRs; instead compare the different institutions' monthly payments for the same amount over the same period.

• A flat interest rate is the constant rate on which most personal loans are made. These don't change so if you borrow £500 over two year's at 12% flat rate, tharis 2 x 12% ie 24% of £500 and that means interest charges amount to £120 which are added to the loan and divided by 24 months which equals a total cost of £620.

• Read the small print. It's boring, it can ruin your eyesight, but it is vital to understand and ask questions when in doubt and confused.

• The golden rule is borrow the least amount over the shortest period and always find out what the total cost of the loan will be at the end of the repayment period.

LOAN SHARKS: Not recommended. Often they knock on doors, or place shifty little small ads in the local paper offering 'easy terms' until you fall behind with the payments, or discover the 1,000% interest charges. **For:** Instant cash. **Against**: Broken kneecaps if you don't cough up.

FINANCE HOUSES: The small and not so small ad section of your local paper will be peppered with companies seemingly willing to advance you a small fortune. As a licenced credit broker the company will be regulated by the Consumer Credit Acts, which is a good indication that their intentions are honourable and that you are not dealing with a loan shark. As in all loan cases you look at the APR (Annual Percentage Rate) to see how much extra you will be paying each month for the privilege and what you want is the lowest repayments over the shortest period. By shopping around you may find their rates can be very competitive when compared with the usual institutions. The interest they charge on top of the loan is called compound interest, which is added to the amount you owe either weekly, monthly, or yearly. As the finance company is at risk lending their own money they usually need some security (i.e. your house). You may be able to get an unsecured loan, although this is usually for small amounts. Do your sums before agreeing to anything. **For**: Easy to arrange. **Against:** Interest, they may want your house as security!

BANKS: Probably the simplest and cheapest way to borrow money officially. Once the loan is approved (usually within 24 hours) the money is sitting in your account waiting to be spent. Having the cash available instantly means that you could drive a harder bargain and even have some left over to slip into your back pocket. What you should avoid though, is over borrowing because the temptation to take out just a few quid more could be overwhelming. You should also keep the repayment period as short as possible which means that you will end up paying less when the loan is over. Interest rates obviously vary between the banks and often that rate stays the same throughout the life of the loan. Shop around, but you will save time and are more likely to be accepted if you approach your own bank. If you don't have an account then you will have to join one of the many well-known and not so well known banks. Loans are usually processed and either accepted or rejected within 24 hours. Once you have completed the application form all you do is wait for the credit checks, references, account enquiries and county court judgement checks that are being made to see if the bank can accept your loan. Rejection may happen if you get a bad credit reference. All lenders have access to information concerning your credit worthiness and if your application fails you are entitled to know why and have any inaccurate information amended. An extension of your overdraft facility also gives you some flexibility provided you agree it beforehand and don't just spend what you have not got. They can also organise insurance to protect your payments in the event of redundancy, illness, or death, but if you are borrowing a relatively small amount then the extra cost, even if it is just a few pounds, probably won't be worth it. Consider also the possibility of a **BRIDGING LOAN** for the amount you intend to spend. No interest is charged on this type of loan, as the repayment has to be made quickly and in full. This would be appropriate if you want to sell your current car, but can't wait to do it before buying a banger. However, you will need to be able to sell the car fast and be able to convince the bank that this is likely. **For**: Quick, easy. **Against**: Interest charges.

BUILDING SOCIETY: They now provide much the same service as banks, which mean all the facilities up to an insurance backed loan. If you have a normal savings account there are usually several loan schemes that you can talk to the manager about. Increasing the size of your mortgage is an option, but can't really be recommended. 25 years to pay back a few hundred pounds? And what will your banger be worth then? Whatever the loan, whether bank or society always do your calculations, how much am I really paying for all this? **For:** Convenient. **Against:** Interest charges.

MOTORING ORGANISATIONS: The well known motoring organisations periodically link up with a high street lender to offer some sort of deal. In most cases you will have to join the organisation, a good thing for those slavishly following principles of Bangernomics. **For:** Quick, easy. **Against**: Interest charges.

FAMILY/FRIENDS: Not necessarily interest, or hassle free, but it can be quick. Start writing to your long forgotten Aunt Edna now. **For:** Quick but. **Against**: Too much to list.

EMPLOYER: If you need a car to get to work, then why not go cap in hand for an interest free loan from your employer. They want you there on time each morning and, especially if public transport is scarce or you work unsocial hours, why should they object? Present your case strongly and this could be a painless route to new wheels. **For**: Proves you have a job for the length of the interest free loan. **Against**: You have to stay in the same job until the loan is paid off.

WEEKLY/MONTHLY BUDGET: Of course, this is something your mother should have told you, or you should have listened more closely when she explained the principles of double entry bookeeping. It is vital to work out just how much spare cash you have each week/month. This will give you the best indication as to whether you can afford a car and in certain circumstances the finance payments. Only you can decide how much to spend, so use this chart to help you out.

INCOME Salary/Wage £(   ) Bank, Building Society Interest £(   ) Pocket Money £(   )

EXPENDITURE Mortgage/Rent £(   ) Council/Local Tax £(   ) Electricity £(   ) Gas £(   ) Telephone £(   ) Food £(   ) Entertainment£(   ) Transport £(   ) TOTAL__________

WEEKLY/MONTHLY BUDGET

BUYING BUDGET£(   )

Again, enshrine these figures in ink and stick to them.

Now you have the money, or promise of, you can move on to the next stage.

3. HEAD BANGERS: BEFORE YOU BUY

Sit back in the armchair and relax. Bangernomics is not about making an impulse purchase, based purely on the fact that the car is cheap and cheerful. Bangernomics can only be practiced successfully if the student is prepared to consider all the options and think everything through. Preparation is the key to a stress free purchase.

WHEN TO BUY: Believe it or not, the availability and prices of used cars changes on a distinctly seasonal basis. So as well as having your wits, facts and figures about you, don't forget to consult the calendar and tap the barometer.

The fortunes of the new and used car markets are inextricably linked as is proved each 1st August when manufacturers and retailers prey on our snobbery by consulting the alphabet and changing the registration plates. This creates an artificially high demand for new cars and to help pay for them, a huge number of part-exchanges flood into the trade. This situation provides a variety of opportunities for the banger buyer after a bargain. Throughout July the rate of part-exchanges starts to rise quite noticeably, but it doesn't reach a crescendo until August.

In September, the used market strengthens slightly as dealers begin to sort their stocks out. However, the auction houses become noticeably busier throughout

October as the part-exchanges that didn't move off the forecourt are given their last chance at auction. So if you're confident enough to buy it's likely that you can pick up a freshly valeted and prepared motor at a very good price.

For January, read August. This time, buyers want a new car registered in the New Year and once again the part exchanged cars start to clog up the forecourts and small ads. Also the seasonal weather conditions start to have an effect. Apparently, the thought of pulling on long johns and shuffling around cars in the early afternoon twilight is too much for most people. So what with the January part-exchange fever and the subsequent buyer apathy, there's plenty of scope for finding the car you always promised yourself at the price you know you can afford, right through until March. Then, as soon as the clocks go forward and spring officially, rather than actually, arrives, everyone's thoughts turn to forming a new relationship with a car. The longer evenings encourage buyers to stay out late and the marketplace remains very firm up until August when... Well, if life is a circle, so is the car trade, but the seasonal oddities don't end there.

A mild winter can buoy up a usually lacklustre used car market, whilst a long hot summer gets everyone excited about motoring and sales reach record levels along with the temperature.

Certain models sell particularly badly depending on the time of year, and there's a certain logicality about it all. Therefore a convertible reaches a low point just before Christmas and a 4x4 isn't a high priority during the spring and summer months. However, all it takes is a washout summer to send prices and demand down. If you can anticipate and take advantage of the seasonal swings and roundabouts, good. If you need a car tomorrow, don't worry too much about the weather and get on with it.

WHICH BANGER?: There are thousands of bangers for sale at any one time and the choice can be bewildering. The best approach is to draw up a shortlist of no more than four specific makes and models that you think can be lived with. As a result, you will not be distracted by other cars that might be a similar price, but are unsuitable. You may already have your own set of preferences and prejudices, but there are a few simple guidelines, which should keep you out of trouble:

Understand what constitutes the perfect banger (qv Prize Banger).

What types of vehicles are available that fall into the Banger category. * Refer to the Glossary (A-B of Bangers) for a no punches pulled assessment of the particular models.

PRIZE BANGER

A. MADE IN BRITAIN: Such a claim was not much of a recommendation for home-built cars during the 1970s and early 1980s, but when it comes to bangers there is every reason to feel proud of our workmanship. Well-proven mechanicals is only part of the story.

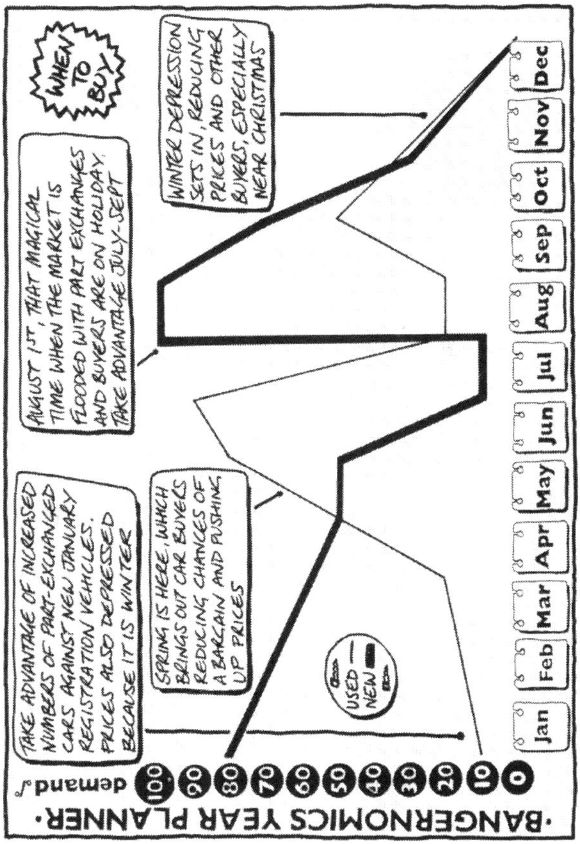
·BANGERNOMICS YEAR PLANNER·
demand
100
90
80
70
60
50
40
30
20
10
0
Jan
Feb
Mar
Apr
May
Jun
Jul
Aug
Sep
Oct
Nov
Dec
WHEN TO BUY
TAKE ADVANTAGE OF INCREASED NUMBERS OF PART-EXCHANGED CARS AGAINST NEW JANUARY REGISTRATION VEHICLES. PRICES ALSO DEPRESSED BECAUSE IT IS WINTER
SPRING IS HERE, WHICH BRINGS OUT CAR BUYERS REDUCING CHANCES OF A BARGAIN AND PUSHING UP PRICES
AUGUST 1ST, THAT MAGICAL TIME WHEN THE MARKET IS FLOODED WITH PART EXCHANGES AND BUYERS ARE ON HOLIDAY. TAKE ADVANTAGE JULY-SEPT
WINTER DEPRESSION SETS IN, REDUCING PRICES AND OTHER BUYERS, ESPECIALLY NEAR CHRISTMAS
USED
NEW

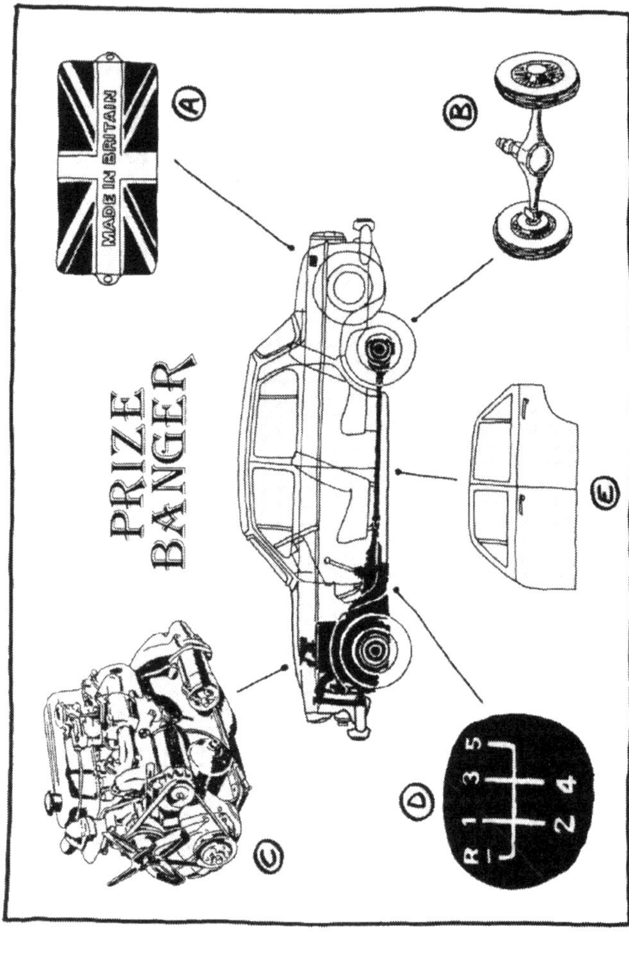

PARTS will be cheap and easy to find, whether you are popping down to the local accessory shop or rummaging around a scrap-yard. As a result INSURANCE is affordable, the parts availability makes brokers smile for once.

B. REAR WHEEL DRIVE: For technophobes out there, all this means is that the rear wheels are the ones powered by the engine, which a simpler and more reliable arrangement than the increasingly common front wheel power layout. It is difficult to avoid front wheel drive in smaller cars, so don't let that put you off, provided it passes your inspection of the car (qv).

C. ENGINE: it is simpler to maintain, cheaper to fix and more likely to be reliable. Overhead camshafts are increasingly common, but can wear out fast and be more expensive to put right; again it is a case of checking the car you want — carefully. Petrol, which is fed to the engine by a carburettor, is easier for the DIY mechanic to look after, whilst petrol injection is more efficient in that it saves fuel, but when it goes wrong then it is a specialist job. However, petrol injection cars are easier to get through the new MOT emissions test, as large carburettored cars are more troublesome. The same goes for electronic ignition, which is maintenance free, until it goes wrong. In addition, go for an engine that is neither under nor over-powered say a minimum of 1.3 litres rising to 2.0 litres, which is suitable for most driving conditions.

D. MANUAL GEARBOX: If you prefer an automatic, only have an automatic driving license, then there is no contest, but when they go wrong it will usually cost double the amount for a manual gearbox. A manual is also slightly more flexible, more fuel efficient and able to limp home on just a few operational gears.

E. FOUR DOORS: Not compulsory, but provides more

flexibility for the owner and may be easier to sell.

WHICH FLAVOUR BANGER?

SMALL HATCHBACKS: Ford Fiesta, Austin Metro, Peugeot 205/104, Vauxhall Nova. *Owner Profile:* 1. Originally purchased as a runaround, playing second fiddle to an executive saloon it will pass through a number of progressively less fastidious owners. 2. District Nurse or OAP who runs this little car until it looks like dropping. *Condition:* Initially the mileage may remain low and bodywork kept clean, but the rot will set in eventually, cosmetically and mechanically it could be very tired if used as first-string transport. *Prices:* Can be rather high especially in built up areas where more people want conveniently small shopping trolleys. Automatics are particularly sought after. *Verdict* Not always ideal for Bangernomics purposes as the demand for good examples outstrips supply and only the cheap dross is left. You will get what you pay for.

SMALL SALOONS: Mini/Austin Allegro. *Owner Profile:* Overworked family hack, or cosseted OAP transport. Likely to pass through a lot of hands before becoming Bangernomics material when it becomes transport for first-time drivers. *Condition:* Rusty, faded paintwork, worn interior but all serviceable. *Prices:* Not as sought after as hatchbacks, so prices can be low, the exception is the good old Mini. *Verdict.* If you don't need the load space of a small hatch, this is just as useful for less outlay.

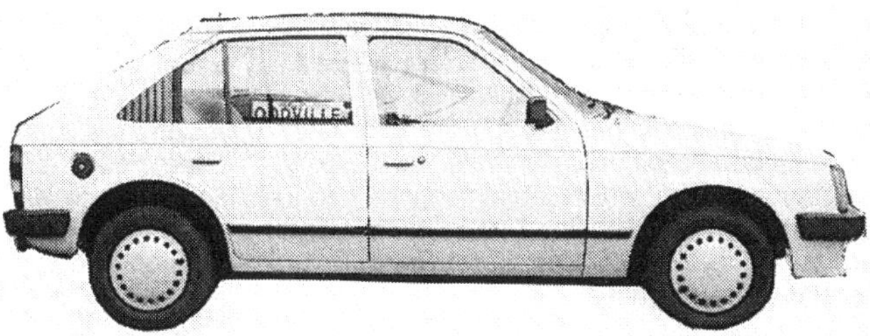

HATCHBACK: Ford Escort, Austin Maestro, Vauxhall Astra. *Owner Profile:* Ultimate family car bought by private owner and consequently looked after. Also likely to be a sample laden travelling salesman special which does huge mileage then disappears into murkier ownership circles and emerges as a banger. *Condition:* Body might look straight, but the mechanicals could be on their last legs. Interior condition could be ragged if it has been used as a surrogate van. *Prices* Stay firm for well known makes such as Escort, but dips for less popular Maestro. Can be excellent value. *Verdict.* Good load swallowers, especially the larger hatches if you don't want a van or estate. Adequate for most uses. Avoid over tired examples

SALOON: Ford Cortina/Sierra, Vauxhall Cavalier. *Owner Profile:* Ultimate travelling rep transport, as a result, the early years consist of high miles and good maintenance, which is good for the car. The opposite happens in later years. *Condition:* These are tough, well proven cars that can stand up to years of abuse. *Prices:* Not surprisingly vary, striking a nice low to middle Bangernomics course which means that you should rarely pay over the odds. *Verdict.* Compromise car, perfect for all circumstances, from long distance to short hops.

ESTATE: Volvo, Sierra, Escort. *Owner Profile:* Bought by the extended family and those with outdoor pursuits, such as towing caravans. They look after them so a one-owner example would be ideal. Picking up a one company-owned estate would also be ideal, but in reality the Bangernomics buyer will come across some pretty ropey models. *Condition:* When they deputise for builders vans, the interior can be shot. Same goes for the suspension, which can take a lot of punishment. *Prices:* Traditionally, estates have always been valued lower than saloon equivalents, although quality family holdalls like the Volvo are one of the exceptions. *Verdict.* Requires careful selection, otherwise the most practical all-rounder that Bangernomics money will buy.

EXECUTIVE: Ford Granada, Vauxhall Carlton, Rover, BMW. *Owner Profile:* Originally bought for company execs, they gradually filter down the system as they depreciate heavily. Favourites of minicab operators. *Condition:* If aforesaid minicab operators have been in charge then the car could be on its last mechanical legs. The problem is that the costs of putting right even minor gremlins can be prohibitively expensive, although minor faults can be lived with. *Prices* The values of the majority of exec-mobiles plummet like stones, not least because of poor mileage per gallon, so they are often dirt-cheap. *Verdict* As a comfy runaround hard to beat, if only using locally and mpg not an issue then could be cheaper to buy and run than a hatchback.

CLASSIC: 1950s-1960s-1970s saloons, Morris Minors. *Owner Profile:* May have had one careful owner or be a multi-owner patched-up rust bucket, or restored by an enthusiast, who knows? *Condition:* From MOT borderline, to nice, honest little runner. *Prices:* There is no need to pay a fortune, tread carefully though as some owners are less realistic and more attached to their cars than others. *Verdict.* The Bangernomics student investigates every possibility and the so-called practical classic movement has bestowed the classic tag on virtually all cars that get past their tenth birthday. Provided you don't pay over the odds these are well proven reliable old buses that should in most instances hold their value and in some rare cases even slightly increase. Don't bank on it though.

VAN: Ford Escort, Austin Maestro, Vauxhall Astra, all car derived vans. *Owner profile:* Bought by companies, driven by psychopaths. Abuse multiplies down the chain of ownership until it looks Bangernomically appropriate. *Condition:* Bought direct from a company with a full history makes sense, but more often Mr. Local Builder is unloading a rusty, mechanically shagged-out heap on the first passer by. Rear load area and doors often rusty and damaged, suspension and engine may also need attention. *Prices:* Can look attractively low but beware the VAT element, which needs to be paid in many cases, especially if you are buying from a company. *Verdict.* Once upon a time vans were brilliantly Bangernomic, being cheap and practical, now they are usually run into the ground so be careful.

SPORTS: Ford XRs, Fiat X19, most Hot Hatches, forget about Ferraris or MGs. *Owner Profile:* Initially the upwardly mobile maniac who passed it on to the progressively younger and more maniacal boy racer *Condition:* The big problem here is a hot hatch that has been wrapped around a lamp-post and then glued together Mechanical items also take a hammering such as engine, clutch and gearbox. *Prices:* When these projectiles start to look Bangernomically attractive it could mean that they have had a chequered history. They might also be cheap because insurance is not. *Verdict* Not ideal material for Bangernomics as a neglected example means big bills.

COUPE: Ford Capri, Opel Manta. *Owner Profile:* Originally young, free and single executive transport, which ends up in the hands of boy racers. *Condition:* Sometimes patched up after a smash, or luridly customised. *Prices:* Ones with an image are priced upwards, some bargains around, but watch the insurance premiums. *Verdict.* Anything that is virtually a saloon car underneath the skin, (eg: a Capri is really a Cortina and a Scirroco in reality is a Golf) bodes well for reliability and parts. Specialist cars should stay with the specialists.

FOUR-WHEEL DRIVE: Land Rover, Suzuki, Subaru. *Owner Profile:* Country dwellers to image obsessed townies. *Condition:* Depends who has owned it. Something that has been down on the farm could be an expensive liability. Suspension takes a bashing and the complex four-wheel drive system can cost plenty to put right. Rust will set in and bodywork is likely to be creased. At the other extreme when used in the soft suburbs it could be in perfect nick. *Prices:* Traditionally stay very high due to fashion trends. If cheap then, there must be a reason. *Verdict* Only appropriate for the Bangernomics student who is going to make some use of the car's capabilities. Otherwise the higher maintenance costs and petrol consumption could be big penalties.

If you are still unsure of what type of vehicle best suits your requirements why not have fun with the YOU WANT/YOU GET chart. It might help and hinder your choice and perhaps cause a few arguments, but its sweeping generalisations could point you in the right direction.

DIESEL: Increasingly, diesel powered cars are becoming more popular and widely available on the used market. So do they make sense. **For**: Economy: 20% better than petrol. Reliability: Fewer parts to go wrong. Residuals: Hold their value better than petrol cars. **Against:** Performance: sluggish compared with petrol versions, although Turbo versions not so bad. Heaviness: mechanical components are heavy duty so power steering essential (standard on many models). Cost: More expensive than petrol versions. Replacement parts are more expensive. **Conclusion***:* For the long distance driver they make economic common sense. Others may not be able to justify the extra initial cost, or buy a good enough diesel within their budget.

BANGER VALUES: Now find out if you can afford any of the cars on your banger short-list. It is simply a case for the moment of looking through the local paper and jotting down prices of the cars you like the sound of. Your library will have lots of papers and magazines that you can read through. This will give you a feel for the local market, which is important as there are regional variations. Also, there is no point in wandering too far for your banger and this valuable piece of research will give you an idea of the age and possibly condition of the one you can afford. If possible, try and reason why one car is priced higher than another. Phone up and find out whether it has had one owner or 32, has no MOT or 12 months worth, private, or dealer sale, all these things make a difference. This is the best way to find out what to pay for bangers. There are also PRICE GUIDES, which can be bought in the newsagent and are the poor man's equivalent of the trade price bibles, yet are very accurate. You will also find abbreviated guides in certain motoring magazines, such as *WHAT CAR?*. *Parker's* and *Motorist Guides* can also give some helpful general pointers as to what you can expect to pay. However, it must be remembered that they are only guides and the condition of the car itself and the deal you can negotiate is the deciding factor Also, they only cover cars up to 10 years old, so it may not cover the banger you had mind.

BANGER SHORTLIST *AVERAGE PRICE*

1 (   ) 2(   ) 3(   ) 4(   )

BANGER BUYING BUDGET £

Which ones can you afford?

INSURANCE Ensure that you are insurable material. Sounds really obvious, but there is no point in going to all the trouble of searching for and buying the banger of your dreams only to discover that no insurance company will touch you, or the car. Legally, you must be insured, the Road Traffic Act says so. If you think its all very Bangernomic to save money by not paying out for it, I am glad to say that you will drummed off the course. Those readers who think they know about insurance can skip read the following few paragraphs as we trawl over the basics. However even know-alls might learn something and even save some money.

There are four levels of insurance:

1. ROAD TRAFFIC ACT: This is the minimum amount of cover demanded by law, which only insures you against injury to another driver but no damage to their property. However, it is rare for insurance companies to consider this type of cover, restricting it to drivers with very poor driving records that they think are bad risks.

2. THIRD PARTY: Which covers the damage you do to someone else and their property, but leaves you to pick up the tab for all your own costs and damage.

3. THIRD PARTY FIRE & THEFT: As the title suggests, if the car gets pinched or spontaneously combusts, you get your money back.

4. COMPREHENSIVE: This pays for most of the damage resulting from an accident for all parties with different policies offering all sorts of additional benefits from tow-in charges to compensation for your injuries.

FACTORS AFFECTING INSURANCE:

A. THE CAR: These are currently rated on a scale of 1-20 and the higher the group the more costly the insurance is going to be. The classification is based partly upon the recommendation of the Repair Research Centre, which assesses cars on their ability to be repaired cheaply and easily. Also tossed into the equation is the value of the car when new, acceleration and top speed. However the older the car gets the cheaper it is to insure. The reason is not because the repairs get any cheaper, there is little real

Photograph reproduced by kind permission of Handmade Films, from the film Withnail and I.

difference, but when the value drops down so low for a costly accident they would rather write the car off and give you its value. So now you know why you need a nice sedate old British car as recommended several pages back.

B. YOUR HOME: The more densely populated the area, the higher the risk of theft and accident. So not surprisingly you could pay 50% more for living in a city as opposed to a rural area. On the same point, if you live in an area that has a bad reputation for theft, or vandalism, insurers are likely to increase your premium unless you garage it.

C. USE: Most policies will allow you to use the car for business without any extra charges, but if it is used primarily for that purpose and lots of other people are allowed to drive it, then the premium will rise.

D. YOU: Believe it or not, you may be an appalling risk, as far as the insurers are concerned. *JOB:* although the vast majority of professions are covered at standard insurance rates, some exception occupations are regarded as very risky. So journalists, scrap metal dealers and market traders often have to pay more, whilst rock musicians and athletes may have trouble getting any cover at all. By contrast most people in steady office jobs such as banks, may even get a reduction on the normal rates. *AGE:* In simple terms, the older you get the cheaper insurance becomes, although at about 70 some insurers will reassess things and may insist on a medical. Below 25 you are regarded as a maniac and clobbered accordingly. From 30 onwards they think you might be responsible and the premiums start to become reasonable. *DRIVING RECORD:* Insurers are very interested in your previous driving history. It should therefore come as no surprise to discover that a few convictions are going to make a difference to your premium. The problem is that each insurer will assess the risk differently. Generally, most won't think too much about the odd speeding conviction, but a drink/drive, dangerous driving, or disqualification under the points system could make a big difference.

POLICY: There are various elements in the policy that can reduce the amount you pay:

EXCESS: This voluntary, compulsory, or combination of the two, amount means that you will pay an agreed sum (e.g. the first £100) of the claim.

NO CLAIMS DISCOUNT: Building up a claims' free driving record means that the cost of the policy will be reduced each year until the discount reaches 60% of the gross premium. Of course, if you have a knock you will lose your entitlement to the bonus. However, in most cases insurers don't take away the whole discount, but just knock it back to say, 40%.

SPECIAL POLICIES: As mentioned before some insurers will offer special rates to certain kinds of drivers, depending on your line of work. There is also some good news if you are a mature driver usually 50+, as many companies operate schemes to cover wrinkly motorists. It is also worth investigating whether the bangers you have on your shortlist could be covered by manufacturers' one-make schemes, or those operated by owners' clubs. If the car is 20 years old and restricted to an annual mileage figure, it may count as a classic and be eligible for special old car cover

Now you know more than enough about insurance, we will briefly come back to the subject in a few chapters time, get some quotes in for yourself on a selection of the bangers. For the moment, just use your usual broker or one of the many some companies who advertise in the newspaper.

INSURANCE QUOTES: £.......£.......£.......£.......

CAR EXPENSES – WEEKLY/MONTHLY

Now let's see if you can afford the car. Try these calculations out on your banger shortlist and see if there is a noticeable cost difference:

I**NSURANCE**: simply divide the quote by 12 or 52 for the weekly or monthly figure.

PETROL: Estimate how many miles you will drive each year (the national average is 12,000, but if you are only going to do local trips then it could be much lower, say 3-6,000). Then find out the average fuel consumption figure for your banger look in the price guides and tables in the back of some car magazines. Divide that figure by the mileage, then multiply that by whatever the horrendous cost of petrol may be £............to arrive at a figure £........that you can then divide by 12 or 52 to arrive at the weekly or monthly figure.

WEEKLY/MONTHLY CAR EXPENSES: (For each car on your shortlist) PETROL £( ) £( ) £( ) £( )
INSURANCE £( ) £( ) £( ) £( )
SERVICING £( ) £( ) £( ) £( )
ROAD TAX £( ) £( ) £( ) £( )
DEPRECIATION£( ) £( ) £( ) £(
MOTORING ORGANISATION MEMBERSHIP

TOTAL £( ) £( ) £( ) £( )

Now compare that figure with the amount you have spare each week/month

PERSONAL ....EXPENSES...........WEEKLY/MONTHLY......

So now you know whether you can afford the banger of your dreams. Remember, stick to your budgets and don't be tempted to overspend.

Finally, if you are already a car owner and plan to replace it with a banger, refer to the BYE BYE BANGER chapter for tips on how to dispose of your car It is in your interests to sell your car first, or at least plan the sale so that the two transactions take place roughly at the same time. You will be in a better position to negotiate with a wad of money in your pocket. If you are buying from a dealer then there is always the possibility of part-exchanging your car. This saves you the hassle of selling, but you will pay a price and not get the full value of your car. However in the BANGER BANTER chapter we will tell you how best to tackle a part-exchange deal.

Still not sure what Banger you want to buy? Here's a chart to help you decide.

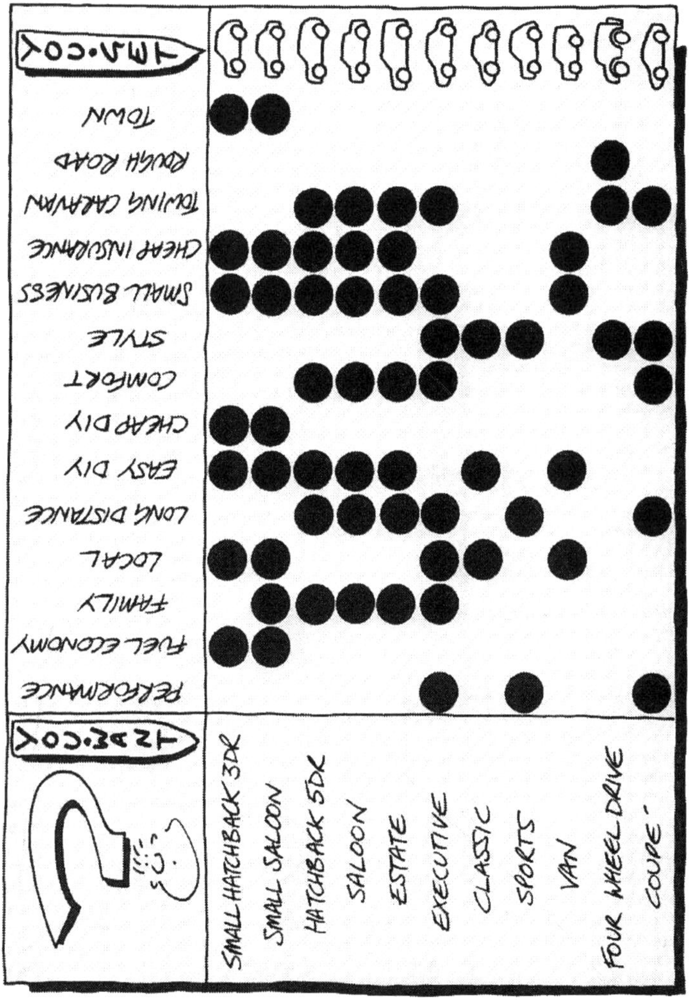
TOWN
ROUGH ROAD
TOWING CARAVAN
CHEAP INSURANCE
SMALL BUSINESS
STYLE
COMFORT
CHEAP DIY
EASY DIY
LONG DISTANCE
LOCAL
FAMILY
FUEL ECONOMY
PERFORMANCE
SMALL HATCHBACK 3DR
SMALL SALOON
HATCHBACK 5DR
SALOON
ESTATE
EXECUTIVE
CLASSIC
SPORTS
VAN
FOUR WHEEL DRIVE
COUPE'

4. BROWSING FOR BANGERS: WHERE TO FIND YOUR BANGER

You know you can afford it, insure it and roughly what you want. Now where is it? For the Bangernomics student, they are everywhere. The best advice is to keep it local. What's the point in tearing all over the country wasting petrol, or travelling costs and time looking at load of old wrecks when you can do that on your doorstep. More importantly, if you want to subsequently bring something to the sellers attention it is better to do it to someone who lives locally and also remind them that you are only a bus ride away.

FRIENDS & FAMILY: There are strong reasons for not buying from your nearest and dearest. What if they sell you a pup for instance? What if they ask too much? That is the risky element in all Bangernomics transactions. However, you have nothing to lose by putting the word about that you are in the market for a car They might know someone, who knows someone else's friend's cousin's mate with the perfect banger. At least you stand a chance of getting a truthful report on the condition and history of the car. Well possibly.

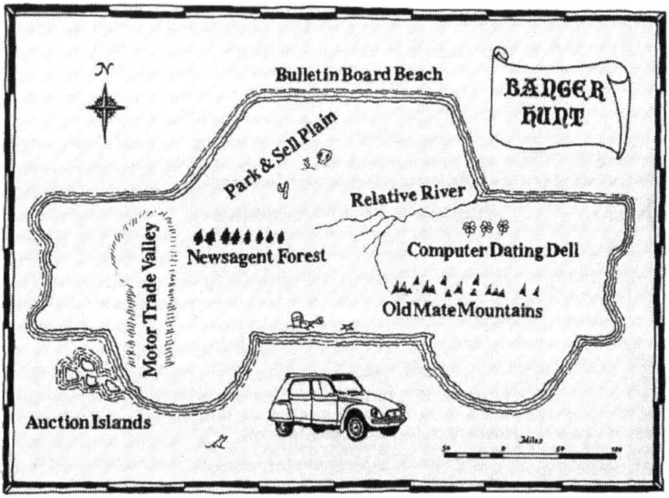

THE LOCAL RAG: Otherwise known as the local press, with nothing to recommend it except the motoring classified section. A good source of mainly private sellers off-loading their wizened motors. You have to act quickly on these ads though, because whether it is a through the letter box freebie, or over the counter gossip sheet, the bargains have been spoken for by the evening of publication. So circle the ads that interest you and ring immediately.

BULLETIN BOARDS: In staff canteens, near the check-out at supermarkets and DIY superstores, leisure centre foyers and in the newsagent's window yes those informative little post cards are everywhere. Amongst the three-piece suites, plumbers and less salubrious services are countless cars in various states of disrepair. This is a popular outlet for motors at the banger end of the market, always worth a look.

PARK & SELL. Increasingly there is a trend for a company to hire a car park, field, or school yard and advertise for people who want to sell their cars to turn up. They pay an entrance fee, park their cars and then let members of the public kick the tyres and, if they are lucky, make a sale. That seems all very well and most events are undoubtedly properly organised, but it can be a magnet for some unsavoury characters and their less than straightforward cars. So long as you regard it as a useful research exercise and see a car you like and check both the seller and car out before buying (qv), then these can be useful events.

COMPUTER DATING: (that's right, there are facilities to put desperate buyers and eager sellers together in the good fashioned lonely hearts fashion. Companies advertise these services widely, inviting you the buyer to ring up and ask the operators to consult their computer database. Give them an idea of the model and price and they will tell you what is available in the vicinity. The buyer pays nothing, but the seller is charged to register the car's details. The problem is that sellers don't always inform the company that their car is sold, which wastes your time. Also, the prices of the cars tend to be high, often to compensate the seller for the cost of registering.

EXCHANGE & MART: This is one of the best established sources of used cars. The only problem is that it operates on a very large regional basis and there are few cars at the banger end of the market. Nevertheless, it is clearly laid out by make, so if you know the model you want, looking through the ads is fast and painless. Recently though, they have introduced an under £2,000 section, so that should be worth a look.

***AUTO TRADER*:** A much more local version of *E & M* but this time with pictures! Although smudgy and small the majority of the ads are accompanied by shots that at least help you identify what it is. There is a bargain section, but it is always worth straining your eyes and checking all of the sections as bangers can turn up in other parts of the magazine.

CAR MAGAZINES: The problem with most car magazines is that almost all are published on a monthly basis. So by the time you get around to enquiring the car has long gone. Also, the glossy nature of the majority of titles means that the motors on offer are usually at the expensive end of the scale. If, however you are in the market for a practical classic the specialist classic magazines offer a good selection of reasonably priced examples both monthly *Classic Car, Classic & Sportscar, Practical Classics, Your Classic, Popular Classics* — and weekly via *Classic Car Weekly* and *Auto Classic.*

THE CAR TRADE

DEALERS: Some of the cheapie car ads may have a (T) which denotes that this is a trade sale (qv on this point). There are many dealers who specialise at the basement end of the market. Often they sell reasonably priced cars that are bargains, at other times dangerous death traps. It is preferable to buy privately as the asking price should be lower, but don't discount the trade sale, because at least you have more legal redress if something should go wrong (qv). The point is that however cheap the car, the dealer will be making a profit of some kind on the deal. Most of the time they will do nothing more than wipe a damp cloth across the paint-work by way of preparation. The thing is that you can save yourself their profit by buying from the same sources, although there are certain Bangernomics risks involved.

AUCTIONS: This is the trade clearing house for unwanted and unloved part exchanges. Dealers who don't want to sell certain types of cars, which have been given in part exchange, go for a quick return by putting them through the auction. See further for details on how to behave at an auction, but bear in mind that in most cases you won't be able to get inside, or drive the car you bid for. Things don't come much riskier than that. (qv BANGERS UNDER THE HAMMER).

NEW CAR SHOWROOMS: You could trace these bangers back one stage to the car showroom by approaching the sales manner of such a concern. The advantage to them is that the car is sold more quickly, without paying an auctioneer's commission. For your part of the bargain you need to make a decision quickly, make the payment in cash and undertake to buy 'as seen' with no comebacks. Not for those who lack confidence. But if you think you can handle the pressure, write or telephone the sales manager, explain who you are and most importantly the amount you want to spend. If you are not too fussy about the type of banger then you will get a faster and more sympathetic response. The most likely dealerships to go for are those that constantly run £1,000 minimum part exchange offers against new cars. They end up with more bangers than they can cope with.

SMALL AD SHORTHAND: Some car advertisements are at first sight almost impossible to decipher. That is because the seller is saving money by often over-abbreviating the information. New and more obscure abbreviations are invented every day, but these are the essential and most common abbreviations used in the small ads.

A, Alloys, AW, Alloy Wheels • A, Auto Automatic Gearbox • ADM Adjustable Door Mirrors • ABS Anti Skid Braking System • Air, A/C Air Conditioning •

BHP Brake Horse Power • CL Central Locking • E/A, E/Aerial Electric Aerial • EDM, e/m • Electric Door Mirror • ESR, E/sunroof Electric Sunroof • EW, E/windows Electric Windows • Fi Fuel Injection • FSH Full Service History • FS/RS Front/Rear Spoiler • History Service History • HR head/r Head Rests • HRW Heated Rear Window • i, inj, in Fuel Injection • LHD Left Hand Drive • LPT Low Profile Tyres • LSD Limited Slip Differential • Ltd Ed Limited Edition • LWB Long Wheelbase • M, Met, MP Metallic Paint • MF, M/flaps Mudflaps • Mls/mlg Mileage • MPG Miles Per Gallon • MSR Manual Sunroof • 0/D Overdrive • ONO Or Near Offer • PAS/PS/Power Power Steering • PDM Passenger Door Mirror • POA Price on Application • PX, part-ex Part Exchange • R/C Radio Cassette • SR Sunroof • T & T Taxed and MOT Tested • T (T) Taxed • TG, Tints Tinted Glass • VGC Very Good Condition • W/W Headlamp Washer, Wide Wheels • 4WD, 4 x 4 Four Wheel Drive

REGISTRATION EXPLANATION

SUFFIX

A February 1963-December 1963 **B** January 1964-December 1964 **C** January 1965-December 1965 **D** January 1966-December **E** 1966 January 1967-July 1967 **F** August 1967-July 1968 **G** August 1968-July 1969 **H** August 1969-July 1970 **J** August 1970-July 1971 **K** August 1971-July 1972 **L** August 1972-July 1973

M August 1973-July 1974 **N** August 1974-July **P** 1975 August 1975-July 1976 R August 1976-July 1977 **S** August 1977-July 1978 **T** August 1978-July 1979 **V** August 1979-July 1980 **W** August 1980-July 1981 **X** August 1981-July 1982 **Y** August 1982-July 1983

PREFIX

A August 1983-July 1984 **B** August 1984-July 1985 **C** August 1985-July 1986 **D** August 1986-July 1987 **E** August 1987-July 1988 **F** August 1988-July 1989 **G** August 1989-July 1990 **H** August 1990-July 1991 **J** August 1991-July 1992 **K** August 1992-July 1993

Registration suffix letters were first introduced in 1963 and first ran from January to December but since 1967 the period runs from 1 August to 31 July. Prefix letters have been in use since August 1983 when all the letter combinations had been used. Some letters were excluded from the sequence, not because the Department forgot the alphabet, but because some letters look too similar or resembled numbers, which might cause confusion. *Note:* Imported cars were allocated suffix letters to indicate the date of importation, but since August 1983 it is to indicate the first year of use whether here or abroad. Where the origin of a vehicle is in doubt a Q plate is issued with a 'Q' taking the place of year letter Usually this is used on kit cars, imports, or cars that have been written-off as the result of an accident and rebuilt. If in doubt, always contact your local vehicle licensing office (the number can be found in yellow pages).

The first contact with the seller is in almost all cases on the telephone. By asking the right questions you can save yourself a wasted journey and find out exactly what you need to know about the car in the ad, use this script to help you. Have a pencil and paper handy to jot down the replies.

TELEPHONE SCRIPT:

I'm phoning about the car (or) **Could I speak to the person who is selling the car?** If you telephone what appears to be a private seller and the reply comes back, 'which one?' Then more than likely you are speaking to a dealer masquerading as a privateer. They do this because it avoids messy things like tax and also side steps the legal obligations that a proper car dealer has. Also, if there is no (T) at the end of the ad, bogus advertisers will put a Christian name which is code for a particular car, so 'Bob' would be the Escort and 'Ted' the Mini, which means in some circumstances the second opening line would be more appropriate.

Can you confirm that the car is...(repeat the advertisement) Recap the details in the ad, sometimes there are printing errors, or even deliberately misleading come-ons. Is the car an L or a GLX? Check the specification. Then clarify the information and make sure you ask the following:

What is the mileage? Often sellers put something really helpful in the ad like, 'average miles' average compared to what? They also round the numbers down, have more faith in someone who says '78,500' with confidence rather than an uncertain 'about 70,000 I think' If relevant you can qualify that with...

Can that be backed up by any service records or other documents? Clarify two points with one question, namely is the mileage genuine and is there a record of maintenance? Ensure that the history consists of either a service book stamped at the relevant intervals, usually 6,000 and 12,000, or a large number of recent bills and receipts.

When was the car first registered? Sometimes the advertisements can be less than clear on this point. You need to establish that a `B reg' is either 1964, or 1984, or a '1985 car' is a B, C or even a car registered in 1984 but claimed by the owner to have the specification of a 1985 car Confused? You might be.

When does the MOT expire? A bald statement that the car has an MOT is not enough, even the phrase 'long MOT' does not help. If there is just a few months or weeks to run, then perhaps the owner won't think that it will pass. So...

Would you agree to the car being MOT'd before sale? (or) **being examined by an engineer?** As you will discover when you read on, an MOT is the quickest and most effective way of finding out whether the car is safe and roadworthy. Alternatively threaten them with a professional inspection and see what the reaction is. If the seller is reluctant to agree, then perhaps they do not have much confidence in their car .

Are there any warranties or guarantees with the car? Apart from some sort of guarantee that might still be effective if the car was bought from a dealer when purchases of major mechanical and other items such as tyres, exhaust, engines often come with a warranty for a limited period or life of the part concerned.

When does the car tax run out and is that included in the price? Always worth asking, as many sellers whip it off the screen and cash the remaining months in before you collect the car

How long have you owned the car? If it has only been owned for a short period then perhaps the owner is trying to unload it before the car expires completely. So...

Is there anything wrong with it? Is there any serious rust? Has it had any accidents? What is the condition of the interior and bodywork? Don't be afraid to ask these straight questions because they normally get straight replies. If you are timid, don't forget that the seller is on the other end of the telephone, so even if they get upset you won't get a black eye. And anyway, someone who gets so upset obviously has something to hide. Consequently you should be able to judge whether the price being judged is fair.

Have you modified the car in any way? A tuned engine, or major body modifications like boy racer spoilers, can make a big difference to an insurance quote, lifting the car from a low to a much more expensive group.

Have you added any extras? Seat covers, decent radio, or anything else that might make your potential ownership just a touch more comfortable.

How many owners has the car had? Obviously the fewer the better, one careful owner rather than a multitude of careless ones.

Where did you buy it from? It is interesting to know whether they got from within the family, a private sale, a used car lot or auction.

Have you got any proof that the car is yours? or (if a dealer) **Have you got the registration documents?** With dealers, sometimes they are lost, at the DVLC, or still with the original owner. Similarly the private seller may have lost them, but should at least have a bill of sale from the time they purchased the car Some explanations are genuinely innocent, others are not.

Is the car on finance/hire purchase? If the car is on finance the company owns it and that will have to be settled before you buy, there are ways of checking this out as you will discover in later chapters.

Why are you selling? For private sellers only. Do they have a convincing reason? Probably the one question that reveals more about the seller and car than any other.

Will I be able to test-drive the car? There is no point seeing a car if that is all you are allowed to do. You must be allowed to drive it so that the car can be properly assessed. It would be worth asking if the seller's insurance covers you, which you must check when you see the car. Perhaps your existing insurance would allow you to drive, otherwise arrange some temporary cover for the purposes of test driving which should not cost very much.

Errrr... Make your mind up time. Do you like what you hear? Don't rush into it, you can always ring back. If the seller objected to the questioning, or was less than forthcoming about the details, be slightly less enthusiastic about taking the matter further After all if they are genuine and genuinely interested in selling they will want to bend over backwards. Bangernomics is as much about gut feelings as practicalities. If you don't like the sound of the car or the seller, make your excuses and put the phone down.

When can I see the car? On the other hand if you are intrigued by the car, make an appointment. **DO** make sure that it is during daylight hours. The banger in question may not be a pretty sight, but you want to see that for yourself. Shadows and dimly lit garages can hide plenty. **DON'T** agree to someone bringing the car to your house, or meeting in a motorway rest area. That can sometimes be the action of someone who has 'borrowed' a car in order to sell it before the police catch up with them. Allow at least an hour for the inspection when making multiple appointments. If you change your mind, have the courtesy to ring up the seller and make an excuse.

5. BANG SPLUTTER CLANG: IS THE BANGER WORTH BUYING?

Now is the time for Bangernomics students to get their hands dirty. As daunting as it may seem there is no reason to be frightened by the prospect of checking a car out. What must be established is that the vehicle is suitable and sound enough for your purposes; then you can call in the experts. The cost effective Bangernomics way is to have the car MOT'd, but it would be just as appropriate to use a vehicle engineer and the pros and cons will be discussed later You could approach every car with a qualified inspector, but only if you had money to burn, or knew a tame one willing to sacrifice their spare time. That means the initial once over is down to you, but don't worry, even if you are mechanically bewildered.

The Bangernomics procedure is straightforward enough for anyone with the minimum amount of commonsense and reasonable powers of observation to make an informed and accurate decision. You will be told why you are doing something and not simply given a bizarre set of confusing rituals, which could embarrass everyone. As a first step to gaining confidence why not carry out the tests in this chapter on a friend's car? Then you will know how long it takes and how simple the tests are to perform.

BANGER SURVIVAL KIT

A. CLOTHES: Although it won't be a fashion parade, you should dress for the occasion and that means scruffily. Any clothing that might be improved by the odd oil stain would be ideal. If you have sensitive hands, or would rather not get them dirty, don't be afraid to wear gloves. These should be fairly thin don't bother with mittens but rubber gloves allow you to poke around more easily. There is also the psychological factor that if you do resemble a tramp, then the seller really believes you when you plead poverty and offer to pay less than they are asking!

B. RAGS: No, these are not your clothes but a couple of very useful items that may have started life as your underwear. You can use them to kneel on when looking around the car when testing the oil, or just to help you wipe away the grime when trying to get a better look at something.

C. MAGNET: You can still pick them up in toy shops, or why not simply pinch one of those daft things that live on the door of your refrigerator. What a magnet does is stick to metal and drop off any replacement fibreglass panels, filler and be reluctant to cling to rust.

D. TORCH: Helps you get a good look underneath the car in wheel arches, even in the boot and under the bonnet. Make sure it works before you leave.

E. MIRROR: Not absolutely essential, but useful in allowing you to see into awkward places.

F. SCREWDRIVER: This is for poking around with, so it need not be a brand new item.

G. CLIPBOARD: Guaranteed to strike fear into any dodgy seller Even if you don't feel professional, it will at least make you look as though you know what you are doing. With that clipboard you should also have a pen and a copy of, or your version of the check sheets in the BANGLOSSARY. This will take you through the tests in this chapter and help you assess the condition of the car. Remember to take any notes about the car that you made when making the initial telephone enquiry.

H. FRIEND: This item is particularly important because it gives an objective view of the car — 'what a heap, you're not buying that are you?' — and to distract the seller so that you can examine the car unhindered. It may help if the friend knows something about cars, but that is by no means a requirement. They are also useful as a human tape recorder to witness the exaggerations of the seller when they swear on their mother's life that the car is Al, or that they will give you £s off.

I. BANGERNOMICS: Yes that's right, keep this copy of Bangernomics up your sleeve, useful for reference and reminders before doing battle with the seller.

INSURANCE Just a final reminder that when buying privately you must have some insurance cover for the test drive. However, dealers must be covered and if they are not, something is wrong.

THE FIRST FIVE MINUTES

1. EARLY: It is a good idea to get to the seller's residence, or forecourt early. Identify the car and then have a good look around before attracting the seller's attention. You might be able to make an instant decision, usually that it is not what you had in mind, so make your excuses and leave. Alternatively, you might catch the seller trying to kick it into life. Also, ensure that you are there early enough in the day to see the car in the light. An amazing number of people are apparently happy to pay good money for a car they could only feel in the twilight. Don't make this fundamental mistake.

2. TIME: You are the customer and, as we know, customers are always right. The car might be cheap, but that is no reason why you should be rushed around it in five minutes and then asked whether or not you want to buy. Take your time.

3. CAR: The first thing to organise is that the car can be clearly seen. If it is parked on a forecourt, this often means that they are sardined door to door, or a private seller may simply open a garage door and invite you to squeeze around it, which is not good enough. Now the inspection can begin.

4. QUESTIONS: This is a good time to recap the questions you asked over the telephone, especially relating to the car's condition. In the flesh, sellers start to `remember' all sorts of fascinating facts about the car, like the time it went for a swim in the Thames. If the life story of the car is changing too rapidly and significantly, then perhaps you should back out before wasting too much time.

5. PAPERWORK: The good news is that you are not going to get your hands dirty straight away. What you need to establish is that the car is what it is claimed to be and that the owner really does own it:

REGISTRATION DOCUMENT V5: This establishes the identity of the owner, or if a dealer, the previous owner, however, the person named as the registered keeper is not necessarily the legal owner. If the car is being sold privately, does the name and address on the document correspond with the person selling it? When buying from a dealer, record the previous owners name and address, you may want to make contact with them later. Note the number of previous owners and the date of the last sale. Perhaps this is a troublesome car being sold in quick succession? As mentioned previously, the fewer the number of owners the ewe likely it has been looked after. Establish the date of registration, is it a 1983 or 1984 `A' registered vehicle. Write down the registration, colour, model description (GTi, GL, etc) engine and chassis or VIN (vehicle identification) numbers so that you can check it against the car itself. Also, hold the document up to the light and you will see a DVLC watermark.

BILL OF SALE: Should the registration document be inconclusive, perhaps the seller has proof that the car was purchased and has the receipt to prove it.

SERVICE HISTORY/BILLS: These are reassuring slips of paper, which prove that the car has been looked after and could back the seller's claim that they recently spent 'a fortune'.

MOT CERTIFICATE: Just because the car has a current test certificate is no guarantee that the car is roadworthy. Confirm that the certificate is valid, the testing centre is usually impressed on the document. Then note the mileage recorded at the time of the last test and compare it with the reading on the mileometer. If someone has been fiddling with the mileage reading that will be obvious, it will also give you an idea of the average mileage, especially if there are several MOT certificates to compare. Also cross-reference the details such as registration, cc (cubic capacity of the engine), date of manufacture and VIN numbers with that on the Registration Document.

ROAD FUND LICENCE: Check that it is valid, it must be if you are going to test-drive the car and also confirm whether it is included in the asking price.

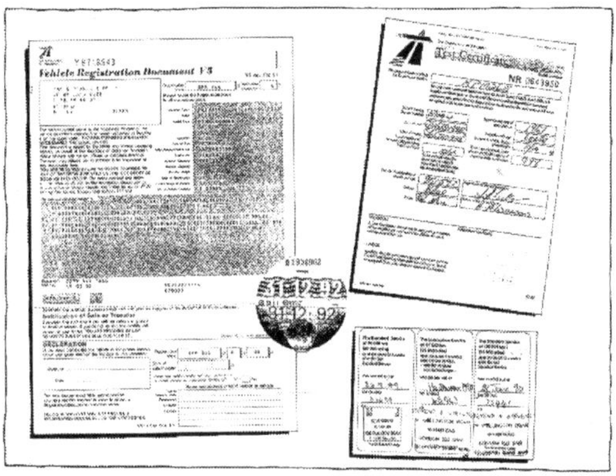

BODYWORK: The idea here is to balance the cost of the car against its condition and suitably for your purposes. If you just want a banger for cheap transportation (which is what this book is mainly about), who cares what it looks like? But you must not buy a dangerous car, or one that will cost a fortune to make serviceable. That is why the preliminary scout around the structure of the car is so crucial, as the costs of rebuilding a rotten body are horrendous.

1. Get the car out into the open, in daylight and in dry weather. Rain does strange things to paintwork and can make it look brighter than it really is. Walk away to a distance of between 12-15ft. Look at the car from the front, rear and both sides. Does the car sit square on the road? If not, is a tyre deflated, or is the road surface uneven? If not this may indicate that the suspension is close to, or has actually collapsed. Alternatively it points to a car that may have been haphazardly bolted back together after an accident. If you feel that the car could be dangerous to drive, then, forget it.

2. Keeping your distance, do the panels look even and

are they damaged? By looking along the sides of the car, from both the front and rear you can see the real condition of the bodywork. Misaligned panels, ripples and general unevenness again suggest cobbling together after an accident. You may also be able to see any differences in the paintwork on the panels, have they been resprayed and why?

3. Get closer and examine those panels. Do any of them look as though they need to be replaced, or simply bear the scars of careless parking? In particular look at the paintwork, If you noticed any contrast in the colours when standing back, inspect them more closely. Generally, although a mono book like this is not ideal for discussion of the condition of paintwork, old paintwork fades and any touch-ups should be obvious. You might see some drips, flecks of fresh paint or a matt surface. A cheap paint job means that there will be

evidence of overspray on the window rubbers, under the wheel arches and on the tyres. The fact that this has been done is not important, but the reasons why it has been done is crucial and that means it is time to grill the seller. Has the car just been tarted up to justify a higher than average price? Perhaps some major rust or damage has been camouflaged. Ask.

4. Open and close all the doors, boot and bonnet, do they fit snugly, or snag on the surrounding metal and do the locks work?

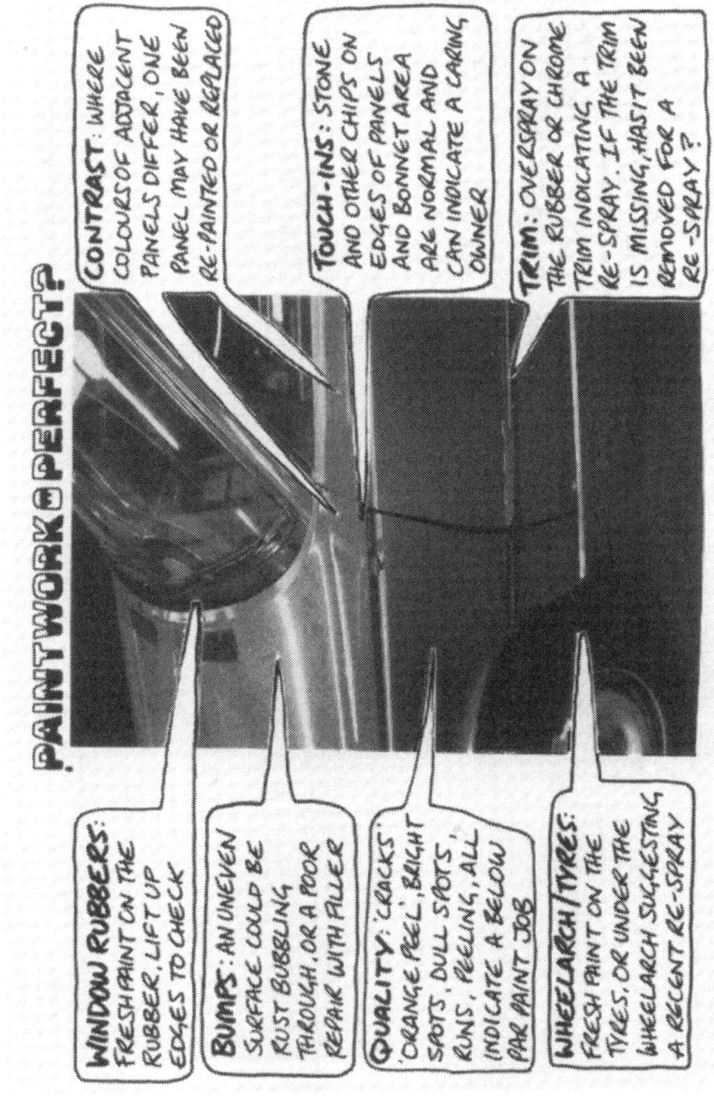

5. Now you are nice and close, this is the time to start looking for rust. This nasty little bug, starts at the

bottom the car where moisture and water collects and then eats its way upwards through the bodywork. What you have to distinguish between is serious and superficial rust. The car may look like a colander but it can still be perfectly legal and safe. Generally, minor rust on the edges of panels is rarely serious, so a rusty bonnet, doors and wings can easily be lived with, or simply repaired. However, structural rust is more serious and results in an MOT failure as the important load bearing areas of the car can no longer take the strain (qv Underside). The problem is that whereas most mechanical parts can be easily and cheaply replaced, serious rot is too uneconomic to repair. If you think that the rust is serious then forget it. Once garages start to replace panels and weld up holes it all starts to get very messy and expensive.

Magnet Will not stick to rusted, or filled areas, so has someone tried to cover up the damage? Wrap a cloth around the magnet so that it does not scratch the paintwork. Press any suspect parts of a panel with your fingers and feel for any give which suggests filler or rust underneath, which may also be accompanied by a 'cracking' sound. You can also find more evidence of rust inside the car, so please refer to that section (qv).

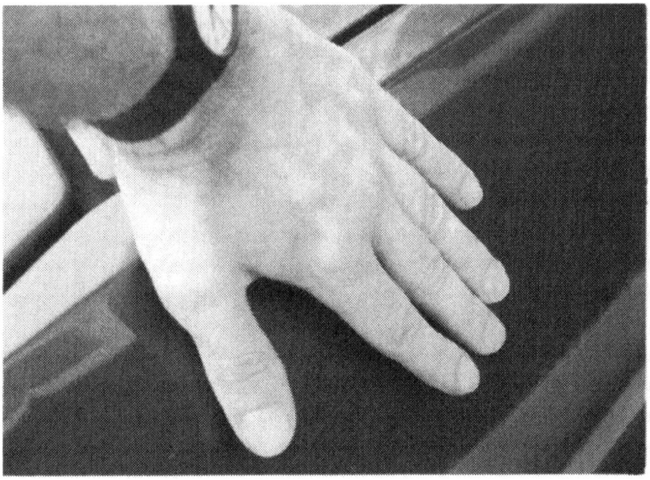

HANDS: These are very sensitive instruments indeed. Running someone tried to cover up the damage? Wrap a cloth around the palm over the bodywork can pick up imperfections and filler magnet so that it does not scratch the paintwork.

SCREWDRIVER: Not to be plunged into the bodywork as the owner is likely to get upset, but useful to clean away mud and muck to see if there is rot underneath. The screwdriver is also excellent when prodding at the underside use it to tap suspect bits of metal. If you hear a pleasing ring, or solid resonate sound then it is solid enough, but a dull thud that absorbs the tap rather than bouncing it off signals rust and filler.

UNDERSIDE It is about time that you got grubby. Of course, the MOT inspector will be able to do a more thorough job, but once again you can save yourself a lot of trouble by eliminating the car at this early stage.

SAFETY: Never get under a car that is supported by a car jack and axle stands. Usually the jacking points are the first to rot and anyway the jack itself is only meant to raise the car high enough to change a wheel. So don't do anything silly like support the axles with bricks. It is perfectly possible to see all that you need by torchlight.

1 SILLS: Firstly look at the sills, if badly corroded they can be expensive to replace, check carefully for any repairs and carry out the tests mentioned earlier.

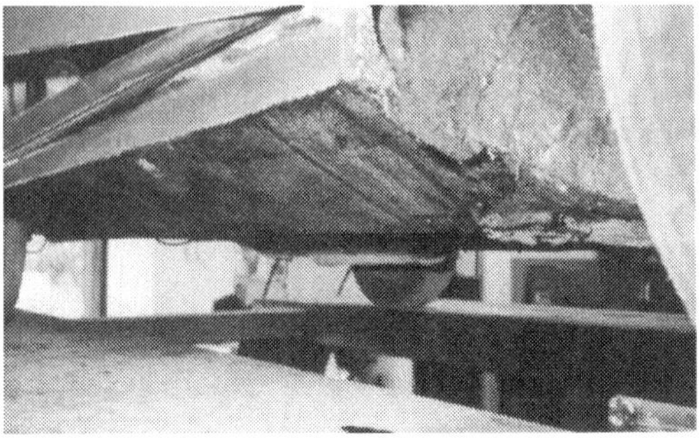

2 SUSPENSION: Now look further underneath and look at the sections, which hold the engine and suspension in place. Do they look twisted or bent? This may indicate an accident, or perhaps there is a clumsily obvious repair?

3. EXHAUST. The exhaust is also worth a look as they rot profusely. There are tests you can perform with the engine running (qv), but for the moment look at the silencer boxes and pipes, are they attached firmly to the car Are they rotten or crudely patched up? Some of these boxes can be replaced individually. If the exhaust looks new, or is made of stainless steel (which last the life of the car), you are in luck and it points to a caring owner.

4. FLOOR: Now look at the floor and the cross sections of metal, again you are searching for rot and signs of inadequate repair usually filler rather than the proper welding job it would need to be safe. Get your magnet under there if you can. Don't be surprised if you see fresh paint and lashings of black sticky underseal place, which are often used to cover up rusty areas. Usually the brown stain of rust will show through.

5. PETROL TANK: At the rear of the car you will find the fuel tank. These can also rot badly, resulting in leaks and leading to fires, so I hope you have put your fag out. Any suspiciously clean no cause for concern, as fuel is an excellent remover of dirt and grime. Follow the petrol line to the front of the car if you can.

6. OIL LEAKS: Look at the differential on rear drive cars and at the engine and gearbox, are they shiny and oily? Minor leaks give, otherwise oil seals may need to be replaced; this can be expensive.

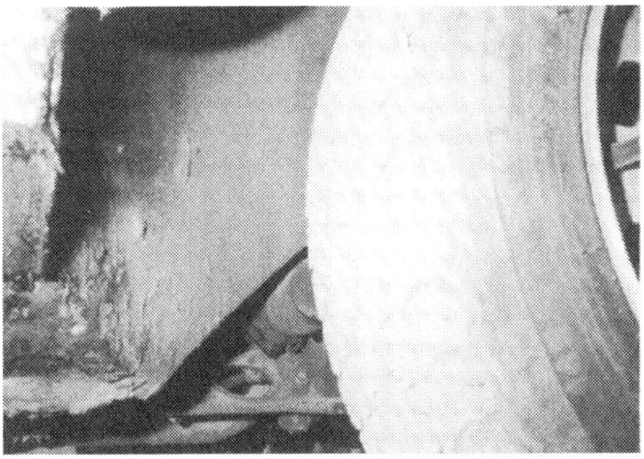

7. WHEEL ARCHES: Whilst you are on ground level peer into the wheel arches which are favourite places for mud and gunge to collect which soon rots the metal. So have a good poke around and be suspicious if there is any freshly applied dirt, which may be covering up some filler.

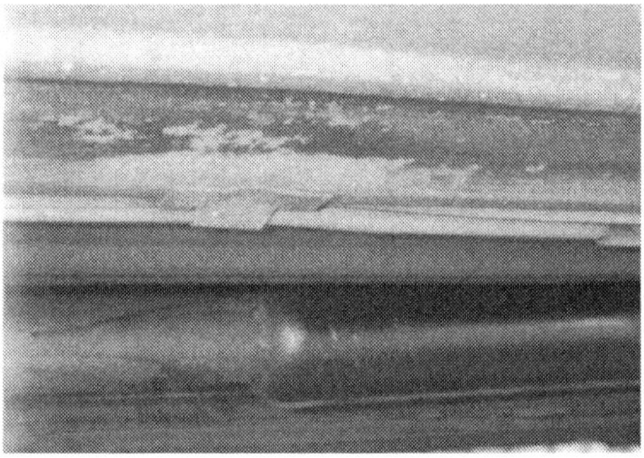

8. BRAKES: The thin pipes leading from the rear wheels to the collect, which soon rots the metal. So have a good poke around and front are the brake pipes. These should be intact, rust free and showing no fluid leaks.

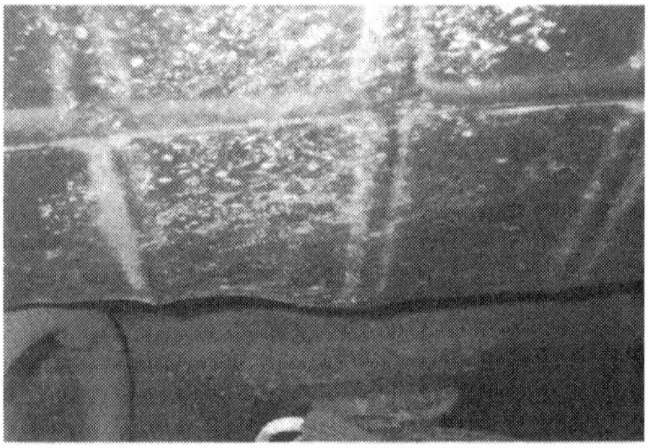

9. TYRES: Whilst you are on the outside, take a good look at the tyres. These are important items because not only is much of the cars' braking and cornering ability lost if the tyres are bald, you can incur driving licence penalty points if they are faulty. Tyre faults include: (a) having incorrect pressures; (b) cuts greater than 25mm or 10% of its section; (c) lumps or bulges; (d) exposed cord or ply; (e) less than 1.6mm of tread over three quarters of the tread width and the tread pattern visible over the remaining quarter. Tyres of different types (radial or cross plies) must not be fitted to the same axle, i.e. opposite each other, whilst radial tyres must not be fitted to the front axle if there are cross plies on the rear. A tyre that wears on the inner or outer surface indicates that the wheel is not vertical or directly in line with the one behind it. A tyre that wears in the middle or has bulges has probably been over inflated. If all that sounds a bit too complicated just remember that cuts, bald spots, bulges and mixtures of tyre types are all bad news and require sorting out. When in doubt don't risk a test drive until that tyre has been replaced. Without getting too complicated, all the information you need is written on the tyre wall itself, telling you its type, the manufacturer and maximum pressure.

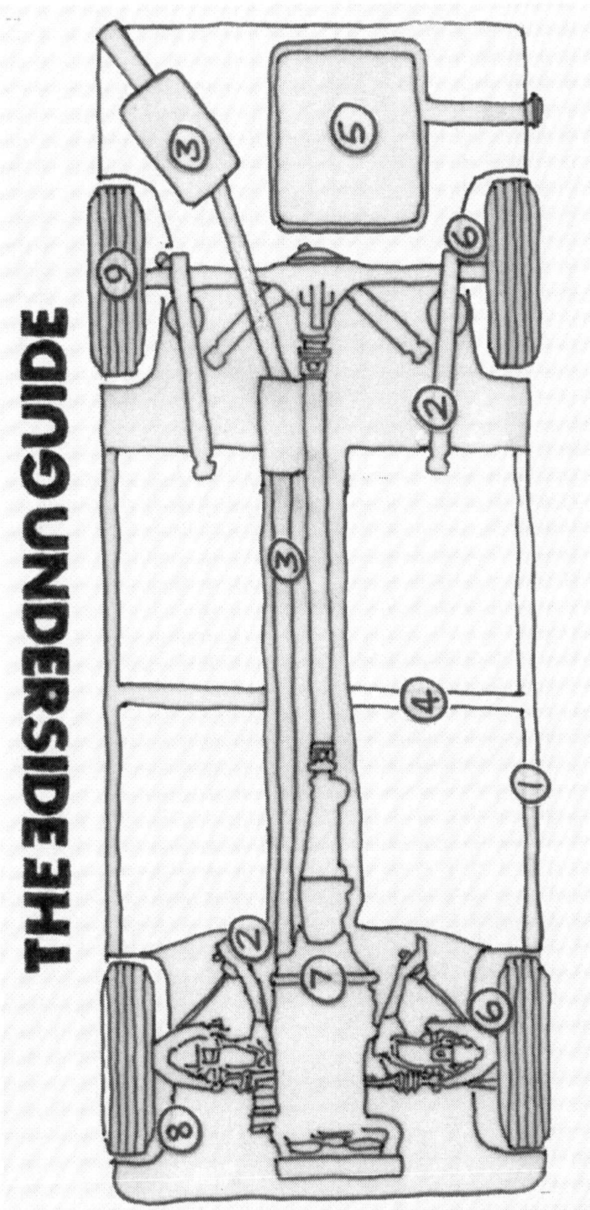
THE UNDERSIDE GUIDE
1
2
2
3
3
4
5
6
6
7
8
9

10. BOOT. Water and moisture can collect in the nooks and crevices of the boot, so move the rubbish out of the way and look for rot, especially the rear suspension mounts. Remove the spare wheel, and look where it has been sitting, a pool of water suggests a leaky boot and probably a rotten floor section. Then check the tyre (qv). If the battery is located here, inspect the area surrounding it, as spilled acid will start the process of corrosion. Is there a wheel brace and jack for changing a wheel? There should be and if you are very lucky perhaps the remnants of a tool kit.

INTERIOR Rather like the appearance of the bodywork, it is a question of what you like and can live with. Some cars are not for the hygienically sensitive, whereas most interiors can be revived with a bit of spit and polish. As ever you pay your money and make that choice.

1 RUST. Try aid lift the carpets; can you see the road through the floor? If badly corroded, then some costly welding may be needed to nurse it through the MOT Otherwise, a musty smell and damp carpets, accompanied with condensation on the windows, also points to a rotting hulk.

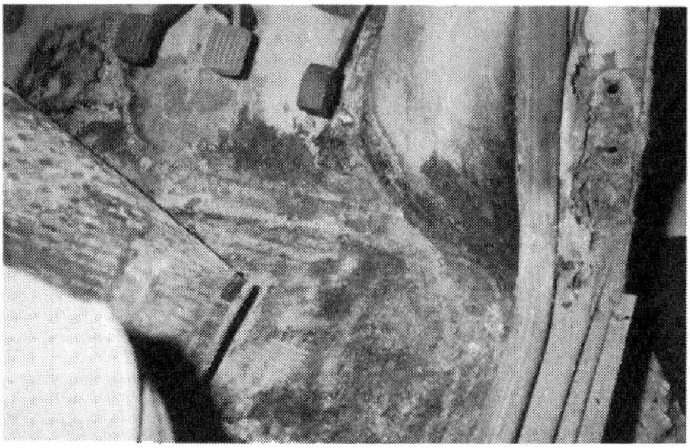

2. WINDOWS: Wind them up and down; do they all work? Then turn your attention to the panes themselves, if security etched with the registration do they match the number plates? If not, why not? A close look at the windscreen from both inside and out could reveal a number of scratches and cracks. If serious it will need to be replaced. Try asking the owner to replace it under their insurance policy.

3. MILEAGE: If a guaranteed mileage is important to you and especially if you are paying extra for the privilege, then the good news is the interior can tell its own fairly conclusive story. A simple respray can hide a multitude of mileage sins, but few sellers go to the trouble of renewing the interior So if the paperwork did not convince you that the mileage was genuine, how does it square with the condition of the interior? Do the seats and carpets look unduly worn. Look at the ignition key, rounded edges point to plenty of use as do worn pedal rubbers, shiny gear knob and a steering wheel with the pattern worn away.

4. SEATS: Remove any covers and assess the seats for comfort and condition. In particular, does the driver's seat adjust? Look at be replaced. Look at the seat belts, the webbing must not be frayed or broken, and the anchor points (where they connect to the body) secure.

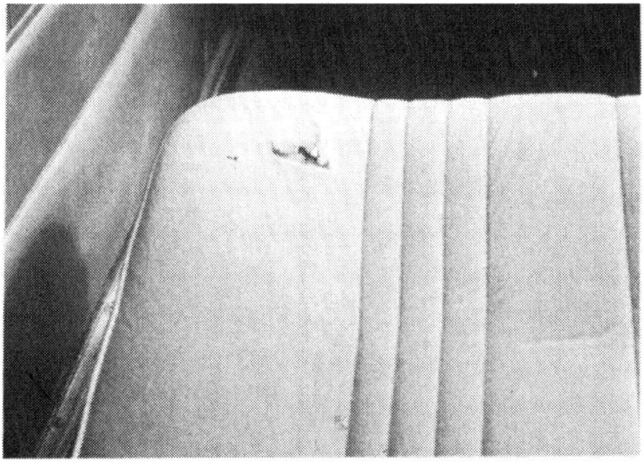

5. LIGHTS: Use your friend positioned outside to check that all I the illuminations are operational, from the indicators and lamps, through to the stop lights.

6. GADGETS: Push, pull and prod all the relevant switches and knobs to make sure that the radio, electric windows, and sunroof all work. Electrical faults can be difficult to remedy and are very annoying. You might be able to live with a silent radio, but a heater is important. It may be something as simple as a blown fuse, then again it might not be.

7. WINDSCREEN WIPERS AND WASHERS: Legally they must work, even if it is not raining, don't be afraid to try them out.

8. **DASHBOARD**: Switch on the ignition and you should find a lot of lights flashing at you and/or gauges indicating something or other Some are more important than others.

OIL. This is usually a red light, often accompanied by an oil can or dipstick symbol. Ideally it comes on with the ignition but goes out within seconds of the engine starting telling you there is enough oil in the system to lubricate the engine. If the light stays on, then there is engine trouble.

IGNITION: This light may have 'IGN' or a battery symbol below it. Not surprisingly it comes on when the ignition is switched on but if it remains alight then either the battery or charging system is faulty.

BRAKE: A red or yellow light for both the foot and hand-brake (B). The handbrake light will go off when disengaged. However if the light flickers then the brake fluid may need to be topped up. Some cars have brake pad wear lights and if they need replacement the light will stay on. When in doubt consult the owner's manual if it still exists, often located in the glovebox or on a shelf.

MECHANICALS This is the part that some Bangernomics students think they might flunk, frightened by the technology. However, there is nothing scary under the bonnet and although the mechanics of the modern motor car are simple, there is no need to know which does what, in order to spot a fault. Once again it is a simple case of keeping eyes and ears open. A few words of warning though, if you have not put your cigarette out, do it now. Also, tuck away any items of loose clothing and tie back long hair if you don't want to become an integral part of a whirling engine. Don't forget that engine gets very hot so keep unprotected fingers away from the exhaust manifold. Note: If the car has been warmed up before you arrive the seller may have something to hide, perhaps it is difficult to start, or it makes strange noises when cold.

1. SUSPENSION/STEERING: The bodywork checks, especially when you looked underneath, might already have revealed any strengths or weaknesses in the suspension. Mainly these relate to corrosion and leaks. By peering into the wheel arches you should be able to see if there is any fluid leaking from the shock absorbers, corroded suspension springs.

Perform the bounce test by pushing down on the wing above each wheel, the car should bounce straight back then settle on the down stroke. If it keeps bouncing, the shock absorbers need replacing.

Grasp the front road wheel and push it forwards and backwards, to see if there is any play in the suspension, especially if you hear any clonks or knocks. (note: certain makes of cars have unique suspension systems such as Citroens, when expert help would be needed). Now rock it against the steering listening for clonks and watching for play in the steering.

Put your hands on the steering wheel and move it slightly, is there any play (no more than an inch), or movement before the road wheels turn? Perfect the double act with your friend to confirm or deny this. If there is play, then the steering may be worn, which the test drive should confirm. This is a possible MOT failure point and may need a specialist to investigate further.

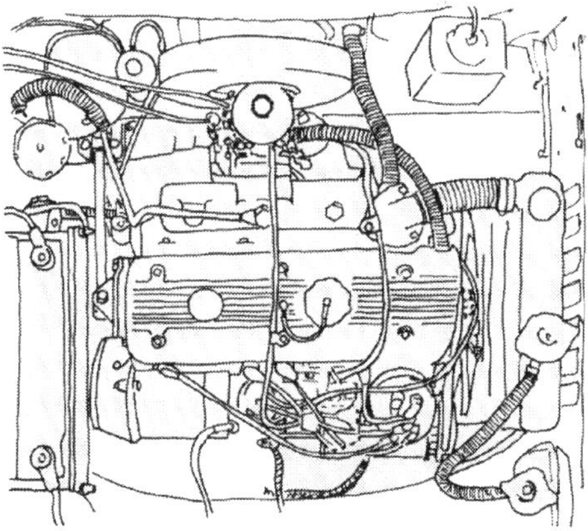

2. UNDER THE BONNET. Don't start the engine yet, first of all try and assess the condition of the compartment.

Anything that looks shiny and new is a bonus, but does it check with what the owner told you has been replaced? If it looks dirty in there, that is the least of your worries. Of course a tidy and well kept area would suggest a caring owner However, dealers and owners who steam clean the engine might well end up with a surface that is clean enough to eat sandwiches off, but it can wash away evidence of neglect. Although there is bound to be some stray oil especially in older cars, a compartment spattered in the stuff could suggest a serious leak.

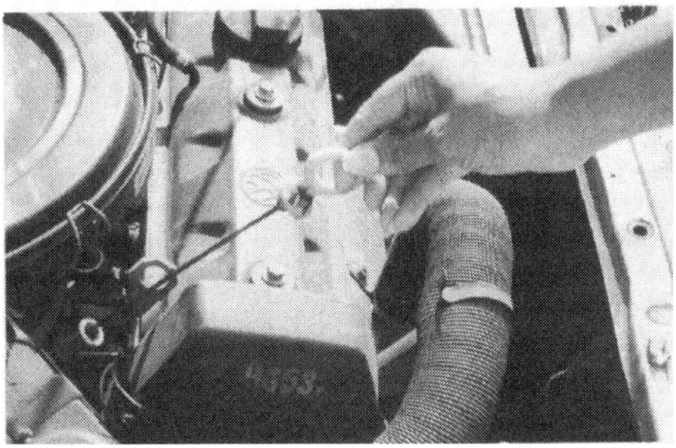

OIL. Look first for leaks around the engine block and gearbox. Get on your knees and look at the engine from underneath, especially the sump, where the oil lives. Move the car or wander back to where it was previously parked, is there an oil slick on the spot? All cars leak to some degree, but worry if it seems to be dripping. Ensuring that the car is on level ground, take out the dipstick wipe it with your cloth, re-insert and withdraw again. The level should be somewhere between the max and min marks. Too low and the car has probably not been cared for. The condition of the oil on the end of the stick tells its own story. New oil is gold and clear like honey. Normally used oil is darker but you should still be able to see the end of the stick. But dirty black oil points to a car that has rarely been serviced. The presence of tiny bubbles suggests that water is starting to mix with the oil, this usually happens when the head gasket blows.

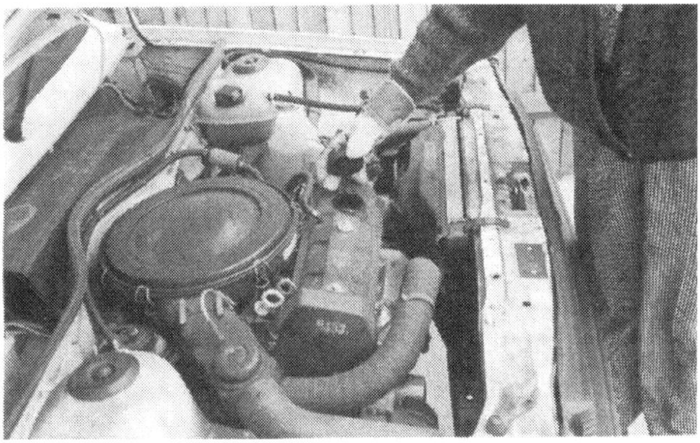

So remove the oil filler cap and look for creamy, mayonnaise like deposits, which suggests water in the oil. Look for the oil filter this is a brightly coloured cylinder, which screws into the side of the engine block. If it looks old and filthy, then again, basic servicing procedures have been skipped.

WATER: Now take off the radiator cap (only to be done with a cold, never a hot engine) and look at the water Once again whitish deposits point to oil and water mixing. Put your fingers in, then rub them together do they feel oily? During winter especially, there should be anti-freeze in the cooling system, which turns the water blue, green or red. Orange water suggests that there is rust in the system, which in turn damages a lot of components. If you can't see any water there may be a leak in the system. Look below to see if there are any drips and look at the condition of the hoses, do they look old perished and cracked? Orange stains are the giveaway to water leaks. Again that points to poor maintenance. This area is interesting if there are any brownish stains, which may even discolour the inside of the bonnet, suggesting that the car has overheated due either to a blocked system, failing water pump, or blown head gasket. It could happen again.

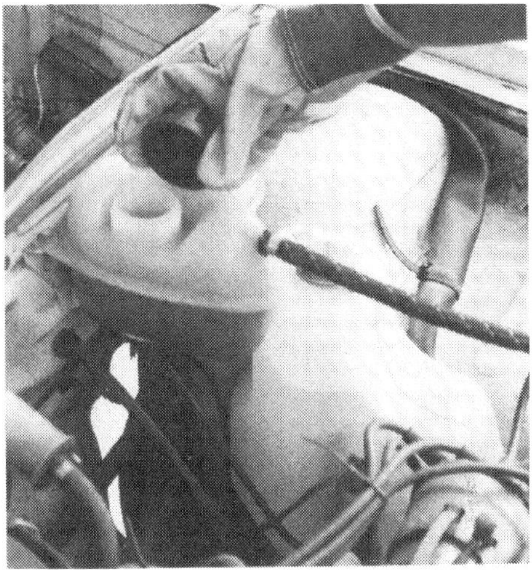

BATTERY. Locate it first most are under the bonnet, some in the boot whilst the odd few can be found under the back seat. Most batteries today are sealed for life, but with any other type, take off the filler caps and check the distilled water level which should be just cover the metal elements. Check the terminals, these corrode and become covered in white powder, which can hamper its effectiveness.

3. ENGINE: At last you can start the car Ask the seller how it is done, there may be some sort of quirk that you are unfamiliar with. Position your tame friend at the rear of the car, ensure the gearbox is in neutral and turn the key. Does it start easily? A slow turnover or several attempts could point to a weak battery. Do all of the warning lights go out? If something is flashing at you find out what it refers too. Listen carefully (qv Nail that noise).

Now get out of the car and ask your friend if they saw any white or dark blue smoke on start-up? An old car is allowed a little puff to begin with otherwise there could be problems.

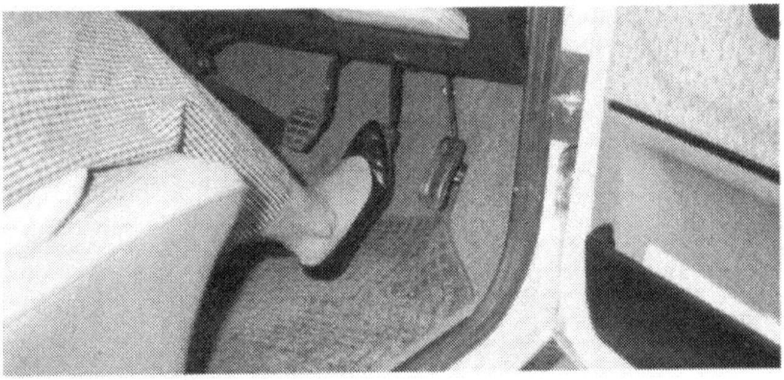

4. BRAKES: Push the brake pedal, it should feel firm and not immediately descend to the floorboards. It's essential to know they are operational before hitting the road and then maybe a wall.

5.GEARBOX initially you can check the gears without the engine running and just make sure that the stick moves smoothly into all of the gears. Then start the car. The stick must not vibrate, which is a sign of wear. More simple tests can be performed on the move.

If you have not already checked the registration document details against the identification plates attached to the engine and body of the car, now is the time to see if they match.

CLUTCH: With the engine running depress the clutch, it should move smoothly and silently, a whirring noise points to a worn release bearing. Engage second gear and the handbrake, increase the revs and let the clutch out. If the car stalls with the revs dropping rapidly then it is alright. Otherwise the clutch is slipping and will have to be replaced.

AUTOMATIC GEARBOX: You can check that the selector moves smoothly into each gear. Then with the engine running and brakes on perform the same checks, the engine should not rev when the changes are made, which should also be silent and smooth. Then, with your foot firmly on the footbrake engage D (Drive) and lightly press the accelerator, then engage R (Reverse). In both cases the car should rise up against the brakes and not stall.

SPECIALIST VEHICLES: WHAT TO LOOK FOR

DIESELS Expect a high mileage. Never buy an ex-taxi as it is likely to be close to collapse. Crankshaft and bearing wear common on tired cars. Never drive a warmed-up car, always start from cold to assess compression, battery etc. Be wary of an oily engine bay and see if there is a leak where the breather pipes connect with the air cleaner this indicates severe wear Also examine the diesel injector pump and pipework for leaks as it operates under very high pressure and crucial to efficient performance, replacement is costly. A service history is very welcome, especially if you are buying up market, i.e. a luxury Turbo.

FOUR-WHEEL DRIVE: Crucial to examine bodywork and underside of vehicle, especially if used off-road. Look for dents, twisted chassis, suspension and corrosion. Find out how the four wheel drive system works and then use it. Listen for noises from front and rear driveshafts and transmission.

TURBOS: Best guarantee of good health is a full service history. Although easy engines to abuse, it is hard to spot a turbo going on the blink. Observe how the owner treats the car (e.g. do they switch off the car immediately after a run: which is bad), listen out for strange noises and assess the car's overall condition.

HISSING, LEAKING WATER PUMP OR HOSE. SIMPLE TO REPAIR, BUT COOLING SYSTEM MAY NEED TO BE CHECKED. BUBBLING – THUMP FROM THIS AREA COULD BE OVERHEATING DUE TO HEAD GASKET AND THERMOSTAT FAILURE. DON'T DRIVE, CHECK WATER LEVEL WHEN COOL
TAPPING AT THE TOP. COULD SIMPLY NEED VALVE ADJUSTMENT. OVERHEAD VALVE CARS COULD HAVE WORN ROCKER ARM OR SHAFT. OVERHEAD CAMSHAFT ENGINES COULD HAVE WORN CAMSHAFTS
WHISTLING LEAK IN CARBURETTOR OR INLET MANIFOLD. ALSO CHUFFING POINTS TO MANIFOLD WITH A PROBLEM.
TICKING FROM MIDDLE. BROKEN PISTON RING. CAN BE A COSTLY GARAGE JOB.
RUMBLING FROM FRONT END, WORN WATER PUMP OR ALTERNATOR BEARING, EASY TO REPLACE, CHECK PARTS PRICES FIRST.
KNOCK FROM MIDDLE WHICH SUBSIDES ONCE ENGINE WARM. NO PROBLEM, CALLED 'PISTON SLAP', PROVIDED OIL LEVEL IS CORRECT.
SCREECHING ESPECIALLY WHEN ENGINE REVVED – SLACK FAN BELT. WHINING – TOO TIGHT, BOTH EASY TO CURE
RUMBLING ROTATING NOISE FROM LOWER PART OF ENGINE, WORN MAIN BEARINGS. ENGINE REBUILD NEEDED.
RATTLING FROM FRONT END INDICATES WORN OR LOOSE TIMING CHAIN. STRAIGHTFORWARD GARAGE JOB ON MOST CARS
HARD KNOCK FROM BOTTOM, WORN BIG END BEARINGS. REBUILD ENGINE.
NAIL THAT NOISE

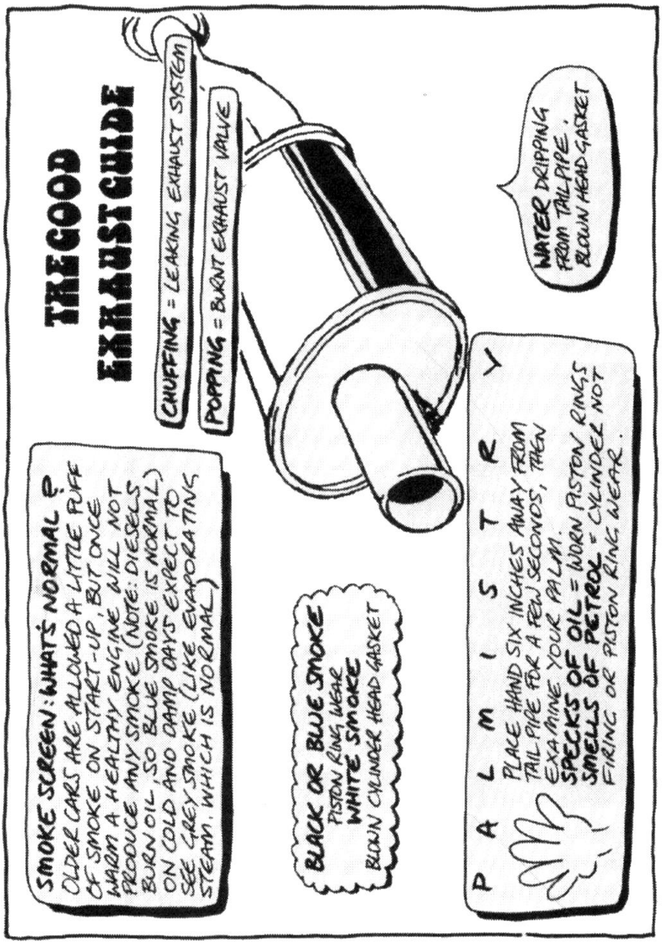

THE TEST DRIVE Having established that the car is safe and legal and provided you have no second thoughts or doubts, make it clear to the seller that you want to take it for a test drive. Not just a spin around the block, but a proper drive of at least 10 miles. If they object because someone else is coming to look at it in five minutes, then you will have to forget it. How else will you know whether the car is reliable and how it performs under normal conditions, or simply whether you are comfortable with it? However if you have had a problem getting insurance at short notice, or the owner is so protective of their car that they will not let you drive, all is not lost. If the owner follows your route and carries out the tests according to your instructions then go along with it. This means that you will have to watch what they do very carefully, as some drivers will attempt to disguise a fault. But if they carry out these tests and you keep your eyes and ears open there is no reason why you should miss a major problem.

The idea is to cover as many different types of motoring conditions as possible, from dual carriageway to town driving. Before moving off and provided the seller agrees, install your friend in the car as an extra pair of eyes an ears, if the radio is on turn it off because it could hide noises and open a window so that you can hear what is going on outside. Don't treat the car as your own and do the seller the courtesy of telling them what you are doing. Remember to keep an eye out for flicking warning lights, or oil pressure and water temperature gauges (when fitted) that seem to be very active.

STAGE 1: Once the car is warmed up, engage the gears and pull sharply away. If there is an audible 'clonk' on rear wheel drive cars this means that the propshaft needs replacing, a major garage job. When accelerating, the car should not jump out of gear otherwise it needs attention. An automatic gearbox should change smoothly and quietly. Compared of course, to the kind of car you are used to, does it feel sluggish? It may be that the engine needs a simple and cheap tune, or it could be that the engine is about to expire. Only a few more tests will tell.

STAGE 2: Drive up a hill (it does not have to be steep) and accelerate; again the car should not jump out of gear Stop and apply the handbrake, then put your foot on the brake and release it slowly. Does it hold?

Perform a hill start and see how the car performs, especially the clutch.

STAGE 3: When descending the hill leave the car in gear and take your foot off the accelerator then provided it is safe, at the bottom, accelerate away. In the rear view mirror you may see a puff of smoke, but not a smokescreen. A smokescreen means a badly worn engine in need of a overhaul.

STAGE 4: Find a quiet road with no one about and at about 30 mph perform an emergency stop. There is no need to slam the brakes on, just apply them firmly. The car should stop in a straight line. If anything dramatic happens, like you start to go excitingly sideways, release the brake and steer your way out of trouble. End the test drive there if you feel unsafe, but bear in mind that brake problems can usually be sorted out fairly easily and quickly by a garage. Should the brakes squeal it could be a simple case of lubrication for a disc brake or

more seriously, worn wheel bearings, whilst drum brakes could be very badly worn.

STAGE 5: Now pull away with the car in second gear, or alternatively, slow right down and then accelerate in fourth. A sound engine will not jerk, stall or make any ominous knocking sounds. Ensuring that you are on a quiet road, bring the speed up to 30 mph and loosen your grip on the steering wheel. If the car pulls noticeably to one side there may be something wrong with the suspension, or an indication that the chassis is twisted after an accident. However there could be a very simple explanation such as incorrect tyre pressures.

STAGE 6: Find some rough ground, or aim the car at some all too common potholes, does the vehicle pitch and wallow, taking some time to settle down? The suspension could be worn, confirming the results of the bounce test, which you should have carried out earlier. Perhaps there is a bang or scrape from beneath the car and the exhaust is loose or touching the road. Don't confuse these sounds with loose tools banging around inside the boot.

STAGE 7: If the car is a front wheel drive vehicle listen carefully when negotiating sharp corners, or stop the car turn the wheel full lock (as far as it will go) and pull away. If you hear a 'clicking' sound, then the driveshafts are likely to be badly worn and can be costly to replace. Any clonks or bangs when cornering suggest worn suspension. If you noticed a grumbling noise from the rear of the car, which may increase, or decrease when negotiating left, or right hand bends, this could be worn wheel bearings.

STAGE 8: On a fast section of road, motorway, or dual carriageway, is there any vibration? This could point to a transmission problem (see below) or if the steering wheel shudders, just a simple matter of balancing the wheels. Put the car into third gear then take your foot off the accelerator and when the car slows, accelerate

again. This 'kangaroo' approach to motoring could reveal weaknesses in the transmission system, so listen out for clonks coming from the driven wheels. Shuddering from the back wheels suggests worn suspension bushes. A grumble from the rear of rear-wheel drive cars indicates worn differential, or wheel bearing.

STAGE 9: On the remainder of the journey back to the seller's home/dealer's forecourt. Listen out for whining, scraping or other unusual noises, which sound out of place. Do the gears engage smoothly or at least easily without grinding? Always remember to put the car into reverse and ensure that it works, so many people forget that you do have to go backwards as well as forwards.

STAGE 10: At the end of the drive, leave the engine running and open the bonnet. Make another check of the warning lights and gauges. By now the engine will be properly warmed up and the oil very thin, so if there are any problems they should now show up. Listen out for noises mentioned before. Also, look around the engine bay, for oil leaks, which will show up after a long run. With a rag in your hand take off the oil filler cap, an excessive amount of smoke indicates wear. DO NOT remove the radiator cap, but look at the hoses, are they leaking?

ANYTHING WRONG? Having previously established that the bodywork is intact and not expensively rotten, leaves the question of the easy, but not always cheap, to replace mechanical items. A slipping clutch, worn gearbox and an excessively noisy engine are all expensive items to put right. Worn suspension, inefficient brakes and uncertain steering are of course highly dangerous. Any of these items should make you think twice before progressing any further If you still like the car but have some doubts about the faults you discovered, then you could ask for a second opinion at a garage. Even if you thought the car seemed very sound, you should still reject it if for any reason you do not feel happy behind the wheel or don't trust the seller Of course, Bangernomics means never having to be fussy, but the priority is a safe car that is cheap to buy and run.

AFTER THE DRIVE If you want the car, what you now need is a short breathing space during which you can make some further checks about the car and seller Of course if you simply walk away from a car it could be sold in the next hour or next day. Place a small 'holding deposit' write a cheque rather than give cash, most sellers won't bother cashing it. Draw up a simple contract (see BANGLOSSARY) and make it clear that the deposit is returnable and make the agreement subject to some condition of your own, such as getting your finances sorted out. You can thrash out a deal now, or leave it until later but do not under any circumstances pay the asking price. Anyway, your checks and second opinions may reveal problems that will enable you to knock the price down even further This deposit is also subject to the car passing an MOT If it fails, they pay, if it passes, you pay and buy it. You can't say fairer than that can you? In reality though, some sellers might take exception to that, which may be because it is an extra half hour of hassle, or perhaps they know it won't pass.

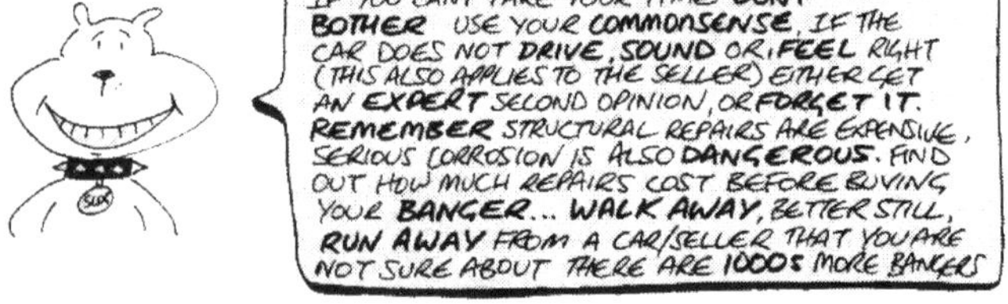

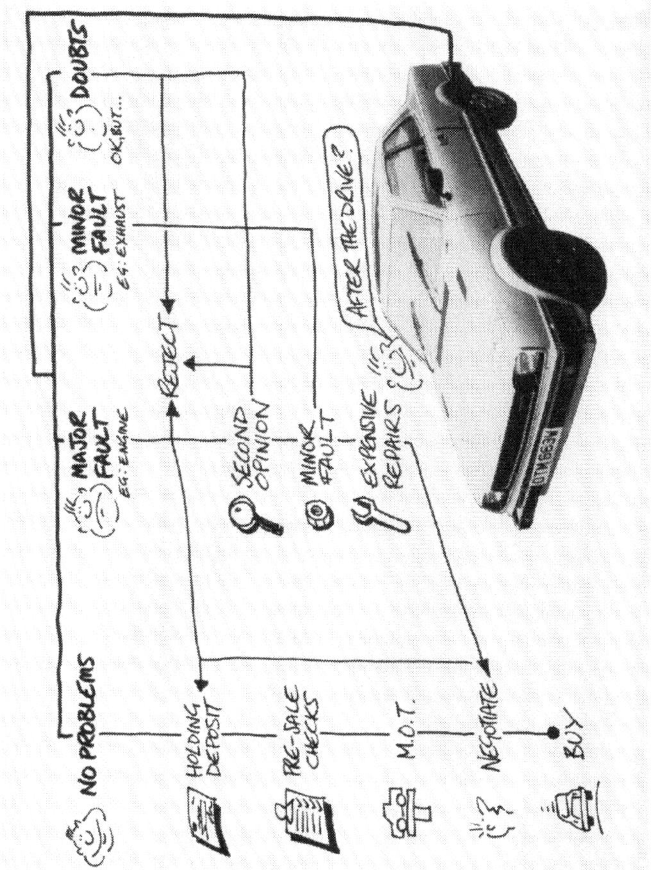
AFTER THE DRIVE?
NO PROBLEMS
MAJOR FAULT
EG: ENGINE
MINOR FAULT
EG: EXHAUST
DOUBTS
OK, BUT...
HOLDING DEPOSIT
PRE-SALE CHECKS
M.O.T.
NEGOTIATE
BUY
REJECT
SECOND OPINION
MINOR FAULT
EXPENSIVE REPAIRS

6. BANGER (BONAO FIDES:
FINAL BANGER CHECKS

A. STOLEN? WRITTEN OFF? STILL ON FINANCE? Now you have that breathing space, get in touch with a useful company known as HPI (Hire Purchase Information). They keep records, which show whether a car is a rebuilt insurance write-off, stolen, or whether it is subject to a finance agreement. At the time of writing your local Citizens Advice Bureau could secure this information for you free of charge, so long as you gave them the registration number HPI could also be contacted direct in certain areas of the country and would charge a fee for this service.

B. REPAIRS Refer back to your check sheet. What items did you spot that would require attention? Telephone your local garage for guidance as to cost and for parts like exhausts and tyres, the local fast fit centre is normally the most competitive place for a quote. A tame mechanic may be able to help you out when it comes to identifying noises, of course they will need to see the car but at least you may get some idea as to whether the problem is likely to be serious or not.

C. PREVIOUS OWNERS If you want to be very thorough and perhaps have a few doubts about the car why not speak to the previous owner? This is especially useful if the owner had very recently sold it. Perhaps there was a problem? Their name and address would be on the registration document, so you should be able to look up their telephone number Contact them at a reasonable hour, apologise for bothering them and ask for their help. Most of us like to be helpful and most owners are happy to either brag or cry about their old cars. Simply ask why they sold it, what was the condition like and what the mileage was.

D. ENGINEER'S TEST A more comprehensive vehicle assessment, including a full road test and an engineer's opinion as to whether the car is worth the asking price, is available from many specialist firms and motoring organisations. The drawbacks are cost and delay (sometimes they can take days to arrange). For the totally mechanically bewildered this is the ultimate in reassurance. In terms of pure Bangernomics, they may be too thorough and actually put you off the car with a list of trivial or minor faults. However for those spending more than the usually low banger budget who mist have their minds put at rest, this is the best investment they are likely to make.

E. MOT It might seem rather pointless to subject your potential banger to a test if it already has a document with a healthy 12 months on it. Perhaps, but a car can easily develop a fault the day after the test and, believe it or not, some MOTs are not entirely genuine. This is your insurance policy in case you missed something (very likely) when looking around the car. The tester has a ramp to see clearly underneath and rolling road to test the brakes. They are also human, and one tester can fail a car whilst another will pass it, purely on the interpretation of an MOT rule, so bear that in mind. Otherwise this is a quick half an hour cheap; expert assessment of the car's roadworthiness. What it will tell you is the condition of the lights, steering, suspension, brakes, tyres, wheels, seat-belts, windscreen wipers and washers, horn, exhaust and bodily structure. And from

November 1991 the exhaust emission level has been tested for the amount of carbon monoxide contained in it. This will give a guide to the state of the engine. In most cases a re-tune should get a car through this test, but it will undoubtedly consign many bangers to the scrap heap. To sum up, an MOT is not perfect or even fool proof, just be aware of its shortcomings. When you are buying from a dealer then 12 months MOT should come with the car automatically. However there can be the implication that when a dealer carries the MOT themselves, or gets a friend to do it, then it may not be worth the paper it is printed on. In some cases this may be true, and if dealer seems less than honest then suggest an MOT testing station of your choice. But in the majority of cases, established dealers and particularly government licensed MOT stations do not want to jeopardise their livelihoods.

F. TAKING THE TEST Book the MOT Make sure that you are there when the car is tested. These days the customers are usually sent to a viewing area or waiting room whilst the operations are carried out. The better stations will summon you over and point out any faults that constitute a failure. If you ask they will also give you an idea of the repair costs. So even though the car may fail, it could be something very minor like a defective light, or windscreen wiper, which would be easy and cheap to fix and still be a perfect Bangernomics prospect. No re-test fee is payable if the car remains on the premises and is repaired there. If it is taken away then half the fee is payable on a re-test within 14 days.

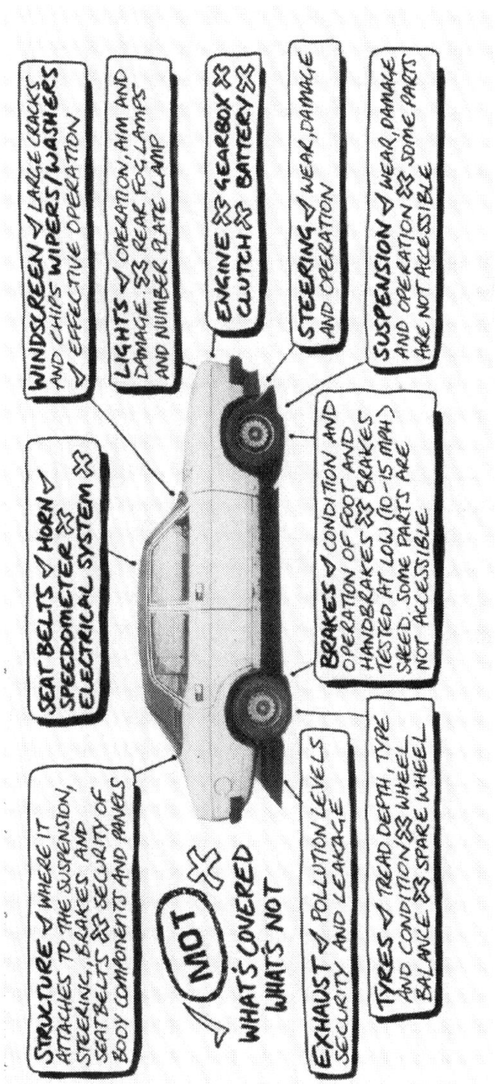
✓ MOT ✗
WHAT'S COVERED
WHAT'S NOT
STRUCTURE ✓ WHERE IT ATTACHES TO THE SUSPENSION, STEERING, BRAKES AND SEATBELTS ✗ SECURITY OF BODY COMPONENTS AND PANELS
SEAT BELTS ✓ HORN ✓ SPEEDOMETER ✗ ELECTRICAL SYSTEM ✗
WINDSCREEN ✓ LARGE CRACKS AND CHIPS WIPERS/WASHERS ✓ EFFECTIVE OPERATION
LIGHTS ✓ OPERATION, AIM AND DAMAGE. ✗ REAR FOG LAMPS AND NUMBER PLATE LAMP
ENGINE ✗ GEARBOX ✗ CLUTCH ✗ BATTERY ✗
STEERING ✓ WEAR, DAMAGE AND OPERATION
SUSPENSION ✓ WEAR, DAMAGE AND OPERATION ✗ SOME PARTS ARE NOT ACCESSIBLE
BRAKES ✓ CONDITION AND OPERATION OF FOOT AND HANDBRAKES ✗ BRAKES TESTED AT LOW (10-15 MPH) SPEED. SOME PARTS ARE NOT ACCESSIBLE
EXHAUST ✓ POLLUTION LEVELS SECURITY AND LEAKAGE
TYRES ✓ TREAD DEPTH, TYPE AND CONDITION ✗ WHEEL BALANCE ✗ SPARE WHEEL

7. BANGER BANTER: DOING THE DEAL AND BEYOND

Everything is negotiable. Even though your banger only amounts to the small change involved in some used car deals, there is no reason why you should not make the price even more buyer friendly.

WHEN TO MAKE AN OFFER Don't plough in too early with a price, bide your time until the last possible moment. Let the seller invest time in you on the test drive and showing you around. Only when you have examined the car and had it MOT'd can you possibly say yes or no, and make the seller understand that to be the case. Because after the protracted process of the test drive and MOT the seller will be desperate to get rid of you at any price! Remember who you are negotiating with. A dealer or trader is more versed in selling techniques, but they are not all high-pressure salesmen. Many have a take it or leave it attitude, itself an effective selling technique. Generally though, stick to the facts and don't be sidetracked or distracted by offers of alternative cars, or deals. Don't be trapped into committing yourself when they ask 'if I could (do this, that or the other) will you buy the car?' Just stick to your original objectives of buying the best car you can afford. The advice is similar with the private seller who might have more experience in the psychology of selling than you imagine. Again, stick to the car and the plain facts, don't listen to sob stories.

WHAT IS IN THE PRICE? The first thing to establish is what is included in the price. This is the time to start reminding the seller of all the promises they made back when you first met. Your 'friend' who came along to witness and restrain your actions should also come in handy in their human tape recorder capacity. So establish, whether the road tax is included, or a tyre is being replaced. When dealing with a dealer a warranty of some kind is likely to be included. These can be very cheap, although seldom worth the paper they are printed on. They should operate like an insurance policy, but even then the small print makes them useless. By all means read what is on offer but in most cases you will see that only a manufacturer's fault rather than wear and tear is covered and that many parts will be excluded, especially electrical items. There may also be a limit on the amount you can claim and a requirement that you pay the first £50-£100 of any claim. This is not worth it in most cases and if it has been lumped in with the price, get it lumped out, but first ask how much it is costing you.

WHAT TO OFFER Of course, the MOT and your assessment of the car could have revealed some faults. Here you have a choice, ask the seller to sort it out, or make an allowance in the price. It depends on your competence to do the job and the accuracy of the quote you may have to rectify the fault. For a dealer it will normally be a cheap or easy matter to deal with and to be honest, do you want the hassle? Now knock a bit off the price. How much is up to you. You are aware of the market by now. Think of a reasonable offer then knock a bit more off. Don't bid stupidly low, or upset the seller By now you should have built up some sort of relationship and if you do get along then that will make the negotiations much easier Remember to keep your humour and wits about you at all times and the supposedly stressful business of doing a deal will seem surprisingly easy.

DOING A DEAL - THE SCRIPT

I'll pay £.... for it. Then stay quiet, don't justify it or um and ahh, be straight and confident and then shut up. Babbling shows weakness. Ideally the seller will come back with a counter offer somewhere in the middle. You can counter that again if the price still seems high. If the seller, plays you at your own game and simply says no, or gives no reaction then...

Thank you for letting me see the car, goodbye. If the seller sincerely wants to do a deal they will stop you disappearing down the road and offer some sort of concession. They may even agree to your first proposal! If there is still room manoeuvre for leaving the road tax on the screen, then compromise, but never exceed your original budget. And that is all there is too it. Be patient, firm but reasonable and always friendly and you should get your way.

PART EXCHANGE Remember that other car in your life? The one that now seems in relatively good condition compared to the one you want to buy. Let's assume that you do not want the hassle of selling it yourself. If you are prepared to part exchange it with a dealer bear in mind that they give the lowest possible figure they can. It will not be a retail price that you could expect from a private sale, but a trade figure, which allows for any potential profit margin when it is resold. Once you understand that, then you are halfway to part exchanging your car What you want to agree then, is 'a figure to change' the net amount of money you will have to pay to get the new car It is worth your while telephoning the dealers who advertise in the local paper for cars and offering your vehicle. That way you can quickly establish a rough trade price and will know what the dealer is really paying. So box around the two figures, understand what the trade value of your car is, but still reduce the asking price of the new car down to arrive at a 'figure to change' that is acceptable to you. Like any buyer a dealer will respond well to a car that looks clean, tidy and well looked after so at least give it a clean.

CONTRACT Once you have thrashed out the deal now is the time to get it all down on paper The details are relatively short and simple, all you need is a brief description of the car the date, the price, if anything has been agreed to be rectified within that amount, the seller's name, address, and signature. It is also nice if you can cover yourself with a few well-chosen clauses, guaranteeing the statements made by the seller (see BANGLOSSARY).

DEPOSITS Although banger buying is essentially a case of the cash hitting the seller's palm and you driving (hopefully) into the sunset, as mentioned before, there are circumstances when you may have to leave a little something so that you can return later on. If there are a few things that the seller has agreed to sort out before collection leave a cheque for a token sum backed up by a cheque guarantee card. Should the seller insist on cash, keep the amount small and get a receipt. Note any sums paid and terms agreed to on the brief contract you draw up together and keep a copy.

INSURANCE Turn back to the Head Bangers chapter I

trust you did research and took notes so that you can now arrange insurance cover Here are some final reminders and tips:

• Look at all the options and talk to as many insurers. So that means high street and Lloyds brokers, specialist car clubs and motor insurance companies. Shop around, but do not choose purely on price, the cover must be suitable to your motoring requirements.

• If possible restrict the use of your car to reduce the cost. Single driver and a limited annual mileage can make a very big difference.

• If you have a garage, use it and tell the insurance company, because in these high crime times, keeping your car out of sight at night can help the premium price.

• Read the policy you are offered very carefully. Undoubtedly it will make your eyes glaze over but if you don't understand any term, get it explained. Ensure that the restrictions you agreed to are included and that you are not liable for any surprise additional charges.

• The voluntary excess provision, can reduce policy charges from 10 to 20 per cent, provided you do not mind paying the first £50-£100, or more of any claim.

PAPERWORK The bottom of the form V5 should be completed and sent to the DVLC so they can issue a new one. Some sellers like to do this themselves as some buyers don't do the necessary and incur driving penalties which are sent to the seller Make sure you get any documentation, especially the MOT Then you can hand over the money.

PAYMENT The easiest way to pay is with cash. If you don't like carrying around bundles of cash then a banker's draft is as good as the folding stuff, but you will be charged for the service. Building society cheques are another alternative if you have such an account and provided the seller is willing to accept it.

COLLECTION A favourite trick for less than honest sellers is to swop the half decent mechanical parts for clapped out items. The new battery that helped to start the car first time could have been replaced by something eking out its last charge. Ditto the tyres from street legal to bald penalty point jobs. So carefully look around for bits and pieces that look newer or even older than before. Only then should you hand over the balance in either cash, or by banker's draft. Also remember to take a simple tool kit, consisting of some spanners and screwdrivers, possibly a jack if the car did not have one. A good idea could be a gallon of petrol. Not surprisingly, many owners siphon off what remains in the tank, leaving you with little chance of reaching the local service station on petrol fumes alone.

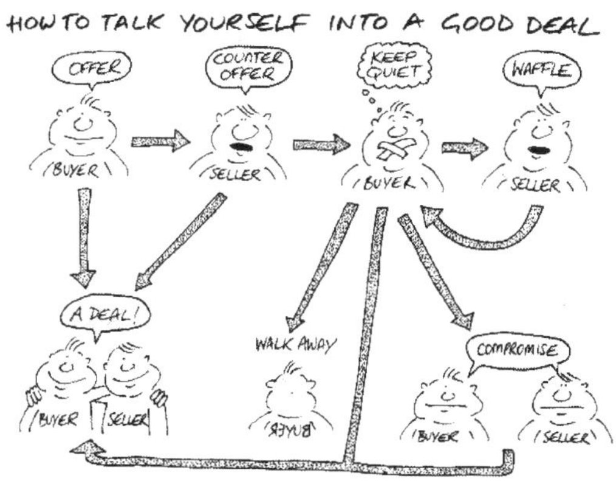

8. BANGED TO RIGHTS: THE LEGAL SIDE

You only get what you pay for, an annoying but accurate saying. It also follows that if you buy a heap, then it was your look out and your mistake. But if someone has misled, or tricked you in some, why shouldn't there be a comeback? This brief legal guide assumes that the seller has not disappeared into thin air but does not allow for the fact that the legal process can move no faster than the average snail.

Let's assume that you have come to an embarrassing, halt within minutes, or weeks, of purchasing your car After slavishly following all the advice in the book, you might be thinking about suing the author But think again, there are several legal options, the problem is though it can be a confusing quagmire. Perhaps Banger of the Bailey our legal expert can clarify the situation.

So I've bought a car from a dealer and it lets me down, what are my rights?

The Sale of Goods Act stipulates that products must be of 'merchantable quality' which, basically, means that they must be fit for their normal purpose. A car would have to be virtually unusable in order for you to reject it on these grounds and claim a full refund. The court would take all the circumstantial factors into account when deciding whether a used car is of merchantable quality, although the standard of reliability would be much less. A court of law would take into account the test of 'reasonableness' could you reasonably expect

12,000 fault free miles from a car that cost £250? Probably not.

But I told the dealer that I wanted the car to tow a caravan and it is totally unsuitable, and that is why it broke down.

If you made your requirements to the dealer clear you have a claim.

What if I let the dealer sort out the car?

If you allow the dealer to put right the fault, however small, you have lost your right to reject the car Alternatively you could allow the repairs but put it in writing that you are accepting them 'Without prejudice' to your existing legal rights.

I remember the dealer hinting that the radio was faulty, can I have it replaced?

There is an obligation on you to examine the car as far as it is reasonably possible, to ensure that it is fit to be used, hence the phrase caveat emptor which means 'buyer beware' Although you won't be expected to dismantle the gearbox, a visual inspection and test drive would be the minimal requirements. If any defects are drawn to your attention, then the merchantable quality proviso is overridden and you are stuck with any faults.

So what other options, legal and otherwise do I have?

There may also be a civil action for breach of contract, if some term of the original agreement has not been met, e.g. a radio fitted, or delivery at a certain time. What you could get is a small amount of compensation.

Alternatively, if the dealer is a member of the Motor Agents Association, they may agree to abide by their conciliation and arbitration procedures.

Should you be unhappy about the decision, then a further appeal lies with the Office of Fair Trading. A dealer will also have obligations under the Road Traffic Act, which means that the car must be roadworthy, unless the car is sold specifically for parts, or breaking.

I paid more because the car had a low mileage, now I'm not so sure.

Dealers are under an obligation to disclose any information they may have about a previous mileage reading, such as when the odometer was replaced. Mileage disclaimers which are often stuck on windscreens are effective provided they are not used to make a representation, e.g. 'the mileage of this car is 10,000, but we cannot guarantee it'

I'm now beginning to wonder about all the other things he said about the car

Dealers must describe the car accurately and must not say that the car is a good runner and if it isn't. Under the Trade Descriptions Act the car must be accurately described to you, whether it is in conversation, or newspaper advertisement, otherwise the Office of Fair Trading can take out a criminal prosecution. Once you rely on a dealer's description it becomes a term of the contract, such as, 'It's definitely a 1978 model' whereas 'I think it's a 1978 model' would not be a term.

What are my rights when I buy privately?

None of the consumer legislation applies. Nevertheless, the Road Traffic Act makes it a criminal offence for anyone to sell an unroad worthy vehicle and there are provisions under the Misrepresentation Act for civil actions involving misdescriptions.

I think he could have been a full time dealer working from home. Tell your local Trading Standards Office, the officers would be very interested to hear from you as they have grounds for prosecution in such cases.

Sounds like a lot of hassle, all this legal stuff.

In some cases, where only small amounts of money are concerned it certainly won't be worth your time and effort. Wherever you can, try and negotiate a settlement. Most dealers are reasonable and have a reputation to think of, so a letter to the owner setting out your problems, politely and concisely, can produce the required results. However if you think that the matter is serious, see a solicitor In the first instance the Citroens Advice Bureau can give you the best indication as to your rights and remedies.

SLOG THE BANGERNOMIC DOG SAYS PRIVATE
SELLERS CAN GET AWAY WITH MURDER!
HOWEVER, THEY CAN'T SELL AN UNROADWORTHY
BANGER REMEMBER IF A DEALER HAS
POINTED OUT A FAULT, YOU CAN'T COMPLAIN
ABOUT IT LATER LISTEN CAREFULLY TO
WHAT A SELLER SAYS ABOUT THE BANGER
TAKE A FRIEND, THEY ARE USEFUL AS A
WITNESS AND CAN STOP YOU MAKING AN
IMPULSE PURCHASE PROMISES MADE BY A
DEALER MUST BE INCLUDED IN THE CONTRACT
FOR PEACE OF MIND BUY FROM AN ESTABLISHED
DEALER, NOT ONE OPERATING IN A PHONE BOOTH
SLOG

9. BANGERS UNDER THE HAMMER: BUYING AT AUCTION

Here it is, perhaps the most risky way to acquire your banger Heads it's a reliable old crock, tails it's a useless old dog. The good news is that a car can be knocked down at almost pocket money prices, which at least offsets some of those worries. In theory it is possible to buy a fairly new car that has had one company owner and is likely to be tired, but well looked after with some life left in it. However that may not always be the case. What you need to buy at auction is confidence and a gambling nature. The Bangernomics course can, however, provide a brief guide to the whys and wherefores of auctions and how to reduce the chances of making a mistake.

WHY CARS GO TO AUCTION

For Bangernomics purposes, entries for the auction come from either the general public or dealers. The dealers dispose of their cars this way for a number of reasons which include: (a) it is an unsuitable part exchange, i.e. a new car showroom would have no interest in reselling a 10-year old heap; (b) the car is too expensive to put right, this may mean mechanically or bodily; and (c) the car may have been stuck on their forecourt without finding a buyer and they would rather have money in the bank than a rusting, depreciating hulk in the showroom.

It is much more rare to find private sellers putting their cars through the auction but they will often do so either because they don't want the hassle of selling privately, or that it has been advertised without finding a seller Again, there may be something wrong with the car that is uneconomic to put right, or won't show up on its brief drive through the auction ring. That is why you need to be so very careful.

Another group of vehicles of interest to the dedicated Bangernomics student are those classed as stolen/recovered and sold by finance or insurance companies. Often they are damaged, but they can be interestingly cheap. Other sources of auction fodder are the fleet and company cars, which are usually no more than two, or three years old, often with a high mileage but backed up by a full service history. Other vehicles come from the public utilities such as Gas Board, or Police, they have lead a hard life but once again maintenance has usually been a priority.

CHOOSING YOUR AUCTION A glance in the local yellow pages should reveal more auctions taking place than you realised. They vary in size from the national chains that sell thousands each week to the tiny independent shifting just a few cars. The Society of Motor Auctions have a code of practice and can recommend members if you have any doubts. Things have improved considerably for the private buyer over the years, with most auction companies very willing to help out. It is best to visit the weekday auctions when there are fewer members of the public about. The weekend and evening sales attract more private buyers who are prepared to pay more for a car and therefore inflate banger prices slightly.

Things happen fast at auction...

12.00 Parked in hall.
12.02 Into the ring.
12.01 Off to join the queue for the ring.
12.03 A quick once over

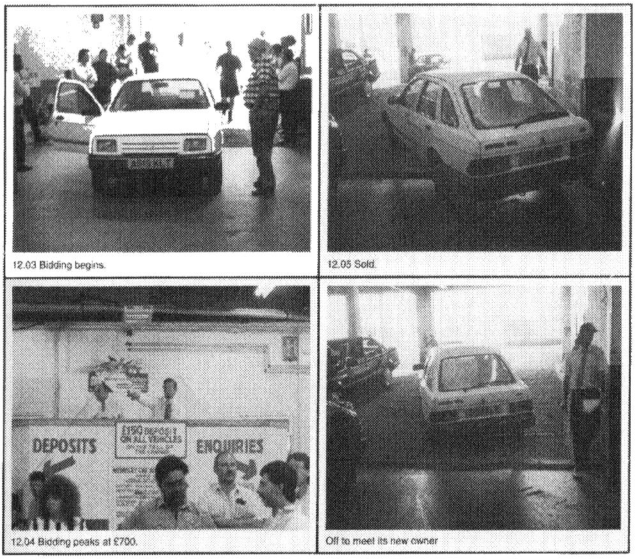

12.03 Bidding begins.

12.05 Sold.

12.04 Bidding peaks at £700.

Off to meet its new owner

YOUR FIRST VISIT This is your research trip. Leave all your money at home and just watch what goes on. It may frighten you off, which might not be a bad thing, especially when you see how quickly it all happens. Two or three visits to one local, or several nearby auction houses, ought to be enough for you to see the types of cars on offer and prices they fetch. Usually our type of bangers kick the auction off to get everyone interested before the expensive stuff arrives. There may even be special 'banger days' or 'evenings', which feature cheaper wrecks. Remember to pick up a copy of the Conditions of Sale and Entry (and any other information) which will make interesting reading and tell you exactly what your rights aren't in some cases! This is very important as the rules vary significantly between the various auction companies. Take a price guide with you so that you can see what sort of money the kind of car you are after is attracting.

AUCTION LANGUAGE The words and phrases used at an auction don't always mean what they say, or say what they mean. Before the sale of the particular lot starts, listen carefully to what the auctioneer says and read the window sticker carefully.

Without reserve This means that the car will be sold to the highest bidder. **Reserve** The minimum the seller will want for the car, not usually revealed before sale, unless it has been entered previously and didn't sell. **As seen** What you see is what you get, including all faults. Some cars fall into this category if they are over a certain age, mileage, or without a reserve which in most cases this means a banger.

All Good also **On description** or **No major mechanical defects** Although this sounds promising, it may not be. It only applies to major components such as engine, gearbox, back axle, steering, suspension and brakes. Other items like instruments, electrics, tyres and trim could all be broken or damaged. Some auctions may offer an insurance-based warranty for a period of up to six months on such a car. **Warranted all good** The auction house guarantees that the car is as described without any major faults. There will usually be a period, from an hour upwards, within which a buyer can reject a vehicle on these grounds. Rarely will a Bangernomics candidate fall into this category. **Specified faults** To say the least, any faults are often seriously underplayed. Phrases like 'worn gearbox' can be translated as meaning 'stuck permanently in second' or 'brakes require attention' can even mean that they don't work at all. It is worth remembering that an auctioneer is acting as an agent for the seller and may only know as much about the car as they have been told.

AUCTION SURVIVAL KIT Cash. The minimum amount required to pay a deposit on the car, read the conditions of entry to find out how much. Sometimes it is 10%, others £50 or £100. Alternatively, some take credit cards. **A friend**. To stop you making any rash bids, or buying something you don't need. **Old clothes**. Make sure you put these on before you get there auction halls are draughty places. **A torch**. **A magnet**. **A price guide**. And finally, if you are thinking of buying something then perhaps a simple tool kit and a gallon of petrol.

BEFORE BIDDING The cars will be parked in long rows. On the windscreen will be a lot number, which gives some indication of the running order of the sale, although this is not always the case. There is also another sticker, which gives some brief information about the car, make, model, mileage, and perhaps a brief engineer's assessment, e.g. 'all good' (but see Auction Language above). However, despite the fact that the assessment says all good, or specified faults you may still be buying 'as seen' under the auction rules if the car falls into that banger category of old age, high miles and no reserve. So be careful and read the rules (Conditions of Entry and Sale).

CHECKING THE CAR To be honest, there isn't much that you can inspect really. For a start the cars are parked closely together which restricts your view. Secondly, in almost all cases, it will be locked. This is a security measure to prevent tampering and even theft, but it is also very frustrating. If you are lucky the seller (often private) may well be nearby, either buffing up the car, or keeping an eye on it. Ask them politely for a look inside and if possible a peek under the bonnet. Perhaps the most important question that you can ask is 'what will it take to buy then?' and you may get a clue to any reserve price (most private sellers put a reserve on their bangers) and what you may have to bid. Otherwise it is a case of using your eyes. Refer to the BANG SPLUTTER CLANG chapter. Except for the test drive you are doing exactly the same checks for obvious defects. Should the car be a stolen/recovered example, look at the ignition, has it been ripped out, along with most of the steering column, and what about the door locks and windows, does it look expensive to put right? Don't pin your hopes on one car it might sell for more than you expected, or once started sound like it has minutes to live. So look at two or three alternatives.

HOW MUCH TO BID? When you have finished, make a note of the lot number the car and the maximum amount you would pay for it based on your experience of visiting the auctions over previous weeks and with reference to the price guides. Don't allocate your entire available budget and bear in mind that the car may require road tax or a new tyre, so take that into account. Also, if you are interested in a van, it will normally be subject to 17.5% VAT on the sale price. This is something that the auctioneer should mention before the sale starts.

AUCTION ACTION The auction begins with the auctioneer outlining the rules of the company concerned. The lot numbers will then be called and the cars are driven into 'the ring' by company employees and parked opposite the auctioneer who briefly describes the car draws attention to any features, then invites bids. The whole process takes about two to three minutes. What you must do is keep an eye on the lot numbers and the cars you are interested in.

1. Before the lot is called, position yourself near to the car your friend can help you with this.

2. Watch the car being started. Does (a) the oil warning light come on and then go off? (b) Does the car start first time? (c) Does it sound healthy, are there any strange thumps, rattles or worrying noises? (d) Is there blue smoke emanating from the exhaust? (e) when the car moves off are there any creaks or groans, the sound of a broken exhaust scraping on the ground?

3. When it is being driven toward the ring it will probably join a short queue, now is your chance for a quick look inside, then ask the driver if they would mind releasing the bonnet. Make a decision now. Is it worth bidding for?

4. If you think it is worth pursuing, position yourself in view of the auctioneer and listen. What they say now is crucial. This is the sales pitch where they will describe the car in glowing terms and stress the good points road tax, automatic transmission, four wheels, etc. But especially listen out for the negative points, 'as seen' faulty gearbox' no documents', etc. If you hear the

phrase `no MOT' forget it.

5. Now is the best time to remind yourself, how much you are prepared to pay and also programme your friend to stop you exceeding that figure should auction fever take over.

6. The auctioneer will suggest an opening bid, which is usually optimistic, 'Do I hear £500 to start?' when £75 would be about right. Don't get sucked into the bidding too early, wait and see what happens.

7. When you do enter the fray raise your arm clearly and confidently and be aware of the amount of each bid, is it being raised in £5, £10, £20 or an even a larger margin? Don't worry if the auctioneer appears to ignore you at first, because they usually concentrate on two bidders at a time.

If you seem to be in the running but can't see who you are bidding against, don't worry; it is not unknown and quite permissible for the auctioneer (within their company rules) to take bids 'off the wall', which really means fictitious bids. They do this to get closer to the reserve price, so be aware that this happens, don't get carried away, or under any circumstances exceed your limit.

If you drop out of the bidding a shake of the head is sufficient to indicate to the auctioneer

10. If you are successful you will be required to pay an immediate deposit some require cash, others will take a personal cheque supported by a guarantee card. However the balance normally has to be paid within a short time period, 24 hours either in cash or by bankers

draft. Dedicated students of Bangernomics will favour cash because a bank will charge you for drawing up a draft. Be aware that you will also be charged an indemnity fee, which insures you against the car being stolen or it still being subject to a hire purchase agreement.

If your bid is successful, but does not exceed the 'reserve' figure set by the seller, then the auctioneers will contact them to find out if they will accept less. In the meantime, don't bid for another motor, or you might end up with two! Should the seller make a counter offer, the decision is yours, but don't under any circumstances exceed your pre-set budget, or pay more than you really think it is worth.

If you have bought a car that came with a trial, read the auction conditions of sale carefully. In most cases you will only have one hour after payment to drive the car and find a serious fault.

If you don't pay the balance of the price within a certain time limit, you will be charged storage fees, so don't delay.

COLLECTING THE CAR Although you should be given the relevant documents, MOT and V5 registration form, a receipt from the auctioneers will be sufficient as proof of ownership if the V5 has been lost. In that case you will need to apply to the DVLC for a new one. Before you drive the car off the auction company's property make sure you are covered by insurance. Driving without insurance would be a very unbangernomic thing to do, as an accident could cost you a fortune and be illegal. Also bear in mind that the car should be roadworthy. One bald tyre, or broken wiper blade is illegal, so fix it first. It is not unknown for the constabulary to lurk in the vicinity and nab the unroadworthy motors as they leave the premises. You should then have it MOT'd as soon as possible for your own peace of mind (even if there are 11 months remaining on the existing certificate.

10. BANGING ABOUT: BANGER 'CARE'

You have safely steered your banger home, it is taxed, insured and freshly MOT'd. What next? Well you could give it a clean. Bangernomics students can now divide themselves into two categories. Leave it as you found it, or beautify and personalise. At the very least you ought to get rid of the more disgusting leftovers belonging to the previous owner banana skins, crisp packets and old passengers that seem reluctant to leave. This can be done quickly, easily and without resorting to expensive specialist 'products.

BODYWORK: All you need is:

1. Some hot soapy water and a sponge to get rid of the grime. If you have some car shampoo use it, otherwise, washing up liquid will do, which, although it does contain very small amounts of salt, bangers can't tell the difference.

2. Rinse off the suds with some clean water and that is it. Now might be a good time to check inside for water leaks, usually through broken windscreen rubbers.

ENGINE: Resist the temptation to clean under the bonnet. Water and engines don't always mix, they have a habit of refusing to start, or suddenly developing intermittent electrical faults. What you can do though, is wipe the grease off the items that you will need to check periodically, (qv BANGER MECHANICS). Use a rag lightly dipped in petrol/paraffin. If the car has not recently been serviced to your knowledge, then a simple oil change will help.

WINDOWS: For your own safety, you ought to be able to see out. The traditional household window cleaners aren't really suitable for cars as they polish the glass and leave a reflective glare. The biggest problem, even after the glass has been washed, is the presence of grease, which smears all over the windscreen. The simple answer is to use the pages of a newspaper, which will soak up the grease. Just rub on and see what muck comes off. Don't forget to rub the paper over the wiper blades too to prevent the grease returning.

INTERIOR: Here a good old-fashioned dustpan and brush, or vacuum cleaner, will do the job. If any of the carpets or mats can be detached from the floor, get them out. Domestic furniture polish will do if you want to buff up plastic seats and dashboards, otherwise a damp cloth will soak up the worst of the dust. If there are some persistent smells, or festering stains, don't be frightened to use well-known brands of disinfectant.

BANGER MECHANICS: If it isn't broken, don't even attempt to fix it. This has to be the Bangernomics watchword when it comes to fiddling about with anything mechanical. Once you start interfering, the fine balance of the car is disturbed and things will start to go wrong, even if you know what you are doing. Some Bangernomics students may care to regard the bonnet as a sealed unit, and simply rely on the experts when trouble strikes. That is one way of approaching it, but of course prevention is better than cure. So a minimum of maintenance and observation should keep everything running smoothly especially if you want to avoid coming to an embarrassing halt. Here is a brief mechanical numbskull's guide to keeping the cogs turning:

WORKSHOP MANUAL: This will save many frustrating hours searching for various items and deciding which spare part is right for the car. These can often be picked up for older cars in junk or charity shops for a few pence. Otherwise there is always the local library.

OIL: At the very least, change the oil unless you have proof (i.e. receipts) that it has been done recently. In simple terms the oil provides lubrication. When it gets old and black, with bits of metal in it, then the liquid is less effective and starts to damage the engine. So changing the oil and oil filter is a good place to start and in most cases finish. If you don't want to get your hands dirty, ask someone else to do it. The more enthusiastic can perform a simple service by referring to the manual.

RECOVERY ORGANISATION: Although the cost might even in some cases exceed the value of your banger, it is essential to invest in this service, especially if you plan to rely on the car. The only exception might be if you restrict the car's adventures to local runs.

TOOL KIT: A motley collection of spanners and screwdrivers are all you really need. The only specialist tool would be a spark plug spanner. Also a good wheel brace is worth having as sometimes the pathetically small spanners supplied with the car have little or no chance of removing the wheel nuts. Other items include: fuel can; cloth adhesive tape; spare bulbs and fuses; tyre pressure gauge; foot pump; battery charger; and, jump leads. Buy good quality items and they will last for the length of your Bangernomics career.

ROUTINE SERVICING: Perform or get someone else to carry out a service in accordance with the workshop manual. Service intervals vary from car to car so an inspection could occur at between 1,500, 3,000, 6,000, 10,000 or 12,000 miles depending on its age and the type of engine. Even if you only cover a tiny mileage annually, you should still check it over at least once a year, otherwise minor faults may develop into major problems.

DON'T TAMPER WITH: brakes, suspension, or steering. For the enthusiastic, though incompetent, apprentice mechanic, steer clear of fiddling with parts that could endanger your health and other people's lives.

WEEKLY CHECKS: What you have to get into is some sort of routine. Put aside 15 minutes on the weekend just to carry out these checks. That means you will be able to quickly spot any deviations from the norm, trace the fault and get it fixed.

WEEKLY MAINTENANCE

1. OIL: This must be kept at the correct level, which is recorded on the dipstick. It is best to test the level when

the engine is cold and when the car is parked on level ground. Remove the dipstick, wipe it with a rag and replace it. Then remove it again and see where the oil level finishes in relation to the max and min marks. If it is hovering around the minimum then it should be topped up. This is done through the filler, which is normally located on top of the engine. If in doubt, check the manual. Buying oil in larger quantities is cheaper, say a five litre can rather than a one litre, and it is also better to buy from an accessory shop or a DIY store.

PROBLEM A: Noticeable increase in oil consumption when checking level and patches of oil when parked. **CAUSE**: Engine oil leaks. SOLUTION: Depends on where leak is; when new gaskets or seals are required then this is a garage job.

PROBLEM B: Increased oil consumption and noticeable blue smoke from the exhaust. CAUSE: Worn engine is burning oil. SOLUTION: Garage may need to fit new valve guide seals, or more seriously new pistons and rings, maybe a new engine!

2. HYDRAULIC FLUID: Both the clutch and brakes rely on specialised fluid to keep them working. Sometimes they are located in plastic bottles and at other times metal pots towards the back of the engine compartment. Refer to your manual for the correct levels. Only a tiny amount of fluid is necessary to maintain the correct level. Do not allow the level to drop as air could enter the system and seriously affect their efficiency. Be careful when handling the fluid as it is highly corrosive both to paintwork and skin.

PROBLEM: Brake fluid level drops rapidly. CAUSE: Leak in system. **SOLUTION:** Don't drive the car and consult a garage.

3. WATER: Some cars have sealed cooling systems, which do not require regular attention, but it is always worth checking the water level as you don't want to overheat. Never check when the engine is hot and, if warm, use a cloth to remove the radiator cap. Refer to manual, but usually the water must be visible and covering a level mark. Plastic expansion tanks are also common and will usually be marked high and low on the side. This is also a good time to check for anti-freeze especially if it is getting towards winter. If there are no greenish traces, buy some from an accessory shop and follow the instructions.

PROBLEM A: Water level drops rapidly. CAUSE: Leaks

from hoses, possibly radiator. SOLUTION: Replace hoses, fit, or repair radiator. Radiator additives to stop leaks are available from accessory shops.

PROBLEM B: Water loss, but also noticeable overheating when temperature gauge high. **CAUSE/SOLUTION**: If there are squeaks and leaks from the water pump then that will need replacing. The radiator could be blocked, or leaking, then stop leak with additive, clean, or replace.

4. WINDSCREEN WIPERS/WASHERS: The plastic bottle is usually found at the side or towards the back of the engine compartment. Make sure that it is full, it is illegal to have an inoperative washer. A hatchback may also have a rear wash and the bottle can normally be found in the boot area. Don't forget to look at the wipers, they must not be perished or frayed.

PROBLEM: Areas were left unwiped, oily streaks which restrict view. CAUSE: Windscreen wipers are perishing. Dirty windscreen. SOLUTION: Replace windscreen wipers and add a small amount of washing up liquid, or special windscreen cleaner to the washer bottle.

5. TYRES: Check their pressures on a petrol forecourt, or with a pencil type gauge. Take readings when the tyres are cool, otherwise the pressure increases and gives a false reading. If you overfill release air by pressing loosely against the valve. Use either a foot-pump, or forecourt inflator Increasingly, however these have to be paid for. Also check the condition of the tyre, are there any tears, splits, bulges. Is the tread legal? Any sign of baldness and it should be replaced without hesitation. Don't forget to check the spare too.

PROBLEM A: Tyre pressure dropping. CAUSE: Leak or slow puncture. SOLUTION: Consult tyre specialist as it may be beyond repair. Always get a second opinion if the fault is not obvious enough to you.

PROBLEM B: Uneven tyre wear, especially on the front wheels. CAUSE: Misaligned brakes, or damaged suspension. SOLUTION: May simply need the tracking corrected which is a cheap and simple garage job. Could be suspension, which is more dangerous, consult a garage.

6. BATTERY: Check the levels, keeping the water just above the tops of the cells, use tap water, or distilled water, which can be bought from accessory shops. Check the terminal connections for tightness and clean away any corrosion either with a brush, or just warm water. A layer of Vaseline around the terminal posts will keep the white corrosion deposits away. Do not forget to check the security of the battery which may be secured in position with a strap, if it falls over, it could start a fire and cause considerable damage to the surrounding area from spilled acid.

PROBLEM: Battery levels drop rapidly. CAUSE: Charging rate high, or the battery plates are faulty, or a cell could be leaking due to a crack in the casing. SOLUTION: In most cases the battery is due for replacement. Talk to a specialist if in doubt, though some leaks can be repaired with a special filler.

7. FAN BELT: This vital little item drives the cooling system and keeps the battery charged. A broken fan belt is one of the most common causes of breakdowns and it can cause considerable inconvenience and worst of all damage. It usually runs around three pulleys (crankshaft, alternator and fan) being located adjacent to the radiator Push and pull the belt where it has the longest connection, the movement should be no more than tin. For adjustment, consult the manual, but the usual method is to slacken the bolts on the alternator and move it until the tension is correct. Now look at the belt for signs of fraying or cracks. If you have any doubts about the condition of belt (which normally lasts about two years) then replace it immediately.

8. HOSES: Like the fan belt, these items can cause lots of trouble if neglected. Check the ones attached to the radiator the top hose is the easiest to see. Are there any obvious water leaks, especially from the ends? If so are the clips holding it in place tight enough? Use a screwdriver Are there any cracks or splits? Now squeeze it, if lots of cracks appear or the rubber feels very soft then it must be replaced. Usually this is a simple job and will save you much grief. Look also at the rubber pipes, which connect to the heater, they should also be in good condition.

PROBLEM A: Water level drops rapidly. CAUSE: Leaks from hoses. SOLUTION: Replace hoses, fit, or repair radiator.

PROBLEM B: Water loss, but also noticeable overheating when temperature gauge high. **CAUSE/SOLUTION**: If the ignition light stays on then the fan belt needs replacing.

9. LIGHTS: An incredible amount of muck is thrown up from the road, which then clings the light lenses. This means other drivers won't see you and in turn you may not be able to see where you are going. So simply wipe them over with a damp cloth or sponge. Periodically check that all the lights work with a friend doing the honours on the outside. Alternatively, if you haven't got any friends then the reflections from shop windows can provide an instant check.

10. GENERAL: Simply make a mental note during your weekly motoring trips of any strange noises, or driving characteristics. Refer back to the BANG, SPLUTTER, CLANG chapter or consult your local garage to identify the problems.

CUTTING RUNNING COSTS

Buying a cheap car is only part of the solution to lowering motor costs. This means that you must continue to think Bangernomically when it comes to running you car Even if you don't feel competent enough, or just can't be bothered to carry out DIY maintenance, that is no excuse for paying through the nose for servicing or parts.

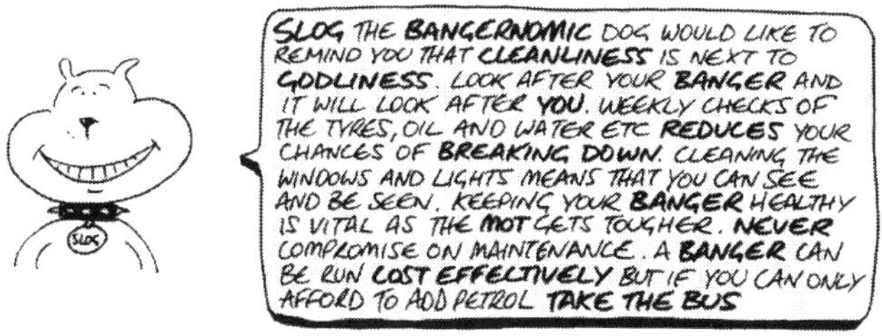

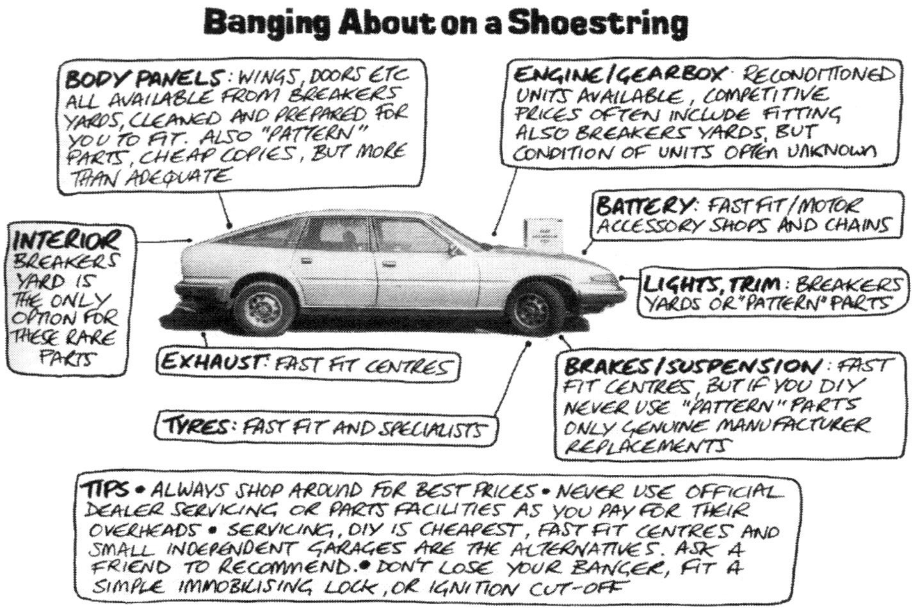
Banging About on a Shoestring
BODY PANELS: WINGS, DOORS ETC ALL AVAILABLE FROM BREAKERS YARDS, CLEANED AND PREPARED FOR YOU TO FIT. ALSO "PATTERN" PARTS, CHEAP COPIES, BUT MORE THAN ADEQUATE
ENGINE/GEARBOX: RECONDITIONED UNITS AVAILABLE, COMPETITIVE PRICES OFTEN INCLUDE FITTING ALSO BREAKERS YARDS, BUT CONDITION OF UNITS OFTEN UNKNOWN
BATTERY: FAST FIT/MOTOR ACCESSORY SHOPS AND CHAINS
INTERIOR BREAKERS YARD IS THE ONLY OPTION FOR THESE RARE PARTS
LIGHTS, TRIM: BREAKERS YARDS OR "PATTERN" PARTS
EXHAUST: FAST FIT CENTRES
BRAKES/SUSPENSION: FAST FIT CENTRES, BUT IF YOU DIY NEVER USE "PATTERN" PARTS ONLY GENUINE MANUFACTURER REPLACEMENTS
TYRES: FAST FIT AND SPECIALISTS
TIPS • ALWAYS SHOP AROUND FOR BEST PRICES • NEVER USE OFFICIAL DEALER SERVICING OR PARTS FACILITIES AS YOU PAY FOR THEIR OVERHEADS • SERVICING, DIY IS CHEAPEST, FAST FIT CENTRES AND SMALL INDEPENDENT GARAGES ARE THE ALTERNATIVES. ASK A FRIEND TO RECOMMEND. • DON'T LOSE YOUR BANGER, FIT A SIMPLE IMMOBILISING LOCK, OR IGNITION CUT-OFF

11. BYE BYE BANGER: BANGER DISPOSAL

When to sell your car is one of life's great imponderables. The exception is when the MOT tester shakes his head and the garage quotes a four-figure sum for helping it through, or the car expires for the 17th time in the space of 200 yards. Alternatively you might need a different kind of car bigger, smaller or faster for your purposes. As you must have gathered Bangernomics shows you how to buy and run a car cost effectively, so to make this a comprehensive guide to selling as well would mean that the book would be twice as thick and twice the price. However turning your car into cash is a very important principle of Bangernomics and, not surprisingly, much of the selling advice is the reverse of the buying advice. I'll assume that you have read the preceding chapters and will only briefly recap where necessary.

CAN YOU AFFORD IT? Doing your sums again, what is the shortfall between what you can expect for your car and the amount you want to spend on the replacement.

WHEN TO SELL

CONDITION: The best time to sell is before the car needs major repairs. Pass the crystal ball. By now you might be able to recognise the telltale sounds of expensive repairs. The principle is the same when buying, serious body repairs are costly, mechanical items can usually be simply replaced. A replacement reconditioned engine may well be cheaper than the hassle of selling and buying an unknown replacement quantity.

TIME: Buying and selling cars is largely a seasonal activity, see the HEAD BANGERS chapter for details. Obviously the right car at the right price will always find a buyer but generally, don't sell in mid-winter or after the August and January new car rushes when there are lots used vehicles about. In high summer everyone is on holiday and you may not get a good response to your advertisement. Spring is a traditionally popular time for people to go shopping for cars.

THE BOTTOM LINE: How much is that doggie in the window? Sooner or later someone will ask and you will have to come up with a firm price for your car that doesn't leave you out of pocket, or the buyer sniggering at your optimism. As rigidly defined areas of uncertainty go, valuing a car takes the dog biscuit. The mistake that most people make is to price their car too high, which frightens buyers away. Despite what they may say, everyone buys on price. Yet it is a fine line between what is and isn't fair because there are plenty of factors that can quickly convert your car from highly desirable to totally unsalable. You know how you fixed a budget for your banger now do the same, consult price guides, look at the local paper then think of a figure and double it. Only joking. More importantly, price your car according to where you are advertising it because it will be compared against other similar vehicles.

PREPARATION: Scruffy cars don't sell, unless they are cheap of course. At the very least, wash it and take a vacuum cleaner to the interior. More expensive cars will benefit from a professional valet and a visit to the garage to repair unsightly damaged panels.

SELLING TO THE CAR TRADE: In the classified section of every local newspaper are dozens of dealers seemingly desperate to buy your old car Problem is they want to pay a 'trade' price. The advantage is that you get a quick decision and instant cash without the hassle of trying to sell privately. If your car has failed its MOT or has died mechanically there are plenty of companies, again in the back of your local paper, or yellow pages who will pay a nominal price and sometimes more for a 'scrap car'.

AUCTIONS: Just as hassle free, but again cars are usually sold at 'trade prices' you will also pay an entry fee and a percentage on the sale price. If you are lucky a bidder could pay more than it is worth.

INCREDIBLE HULK: Potential buyers have sniggered, a dealer offered a derisory sum and the neighbours have got up a petition for its removal. Your pride and joy isn't going anywhere, so what do you do? The simple answer is to chop it to bits. But instead of giving those bits to the dustmen you can actually sell them off individually. The fact is that most bangers are worth more in bits than they are in one piece.

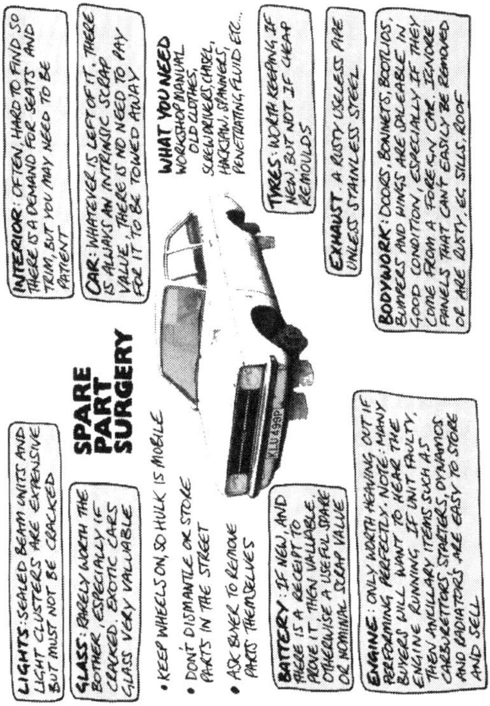

OPTION 1: If you have the space, store the vehicle away and remove the pieces as you sell them.

OPTION 2: Remove the most saleable items from the car and store those away for sale.

WHAT YOU NEED: Most basic tool kits have the necessary items to reduce a vehicle to its constituent parts: spanners; a socket set; pliers; wrenches; cutters; a hammer chisels; hacksaw; releasing fluid/penetrating oil; and don't forget to don old clothes. Power tools can sometimes speed up the process, but they are not strictly necessary. A workshop manual, could also save you inflicting unnecessary damage on a part you want to salvage.

ADVERTISING: Your car may qualify for a free ad in your local paper in the under £50 or £100 section, i.e. 'Breaking Escort, no parts over £50' If your car is a 'classic' or just old, some specialist magazines may give you free publicity.

LEGALLY: If you are selling the whole car for parts, make this clear on the receipt, otherwise you could be prosecuted for selling an unroadworthy vehicle 'Sold as seen for parts'

CLASSIFIED ADVERTISING: There is an art to writing effective motoring classified ads. It is crucial to get the wording right, otherwise you won't get a response and if you do, it won't necessarily be the right sort. The purpose of the advertisement is not so much to fully describe the car as to get interested parties to contact you and buy it. The 10 most important elements in the perfect motoring classified ad are:

1. Year and registered letter: Saying that you have a 1987 car is not good enough because it could of course be a 1987 on a 'D' or 'E' plate. The later plate will justify a higher price and also differentiate between models that may have different specifications. Never underestimate the buyer's knowledge. Providing this sort of specific information is all part of the process of eliminating those who don't want your car and qualifying those who do.

2. Service history: If you've got it, flaunt it. `FSH' (full service history) is an accepted abbreviation. But even if you have no records prior to your period of ownership, there is nothing wrong with stating 'history' as something is clearly better than nothing at all. 'Some history' is a meaningless phrase and is only used by those who are desperate for credibility having stumbled across some dusty old MOT certificates.

3. Mileage: Be specific if the car has a low mileage then say how many; buyers like to know these things. Phrases like low miles, or high miles are just wasting space and money. Ditto the addition of 'genuine' as that is the least that the buyer will expect. The mileage issue becomes less of an issue as the car gets older and can be dropped if you are flogging a banger unless it is 'genuinely' very low.

4. Model: This is vitally important in a marketplace that places such emphasis on the often negligible difference between an L, GL, and GLX. It also justifies a sometimes substantial price differential between two similarly described cars. Again, don't waste words by saying Vauxhall Astra as everyone who is looking for an Astra, knows that Vauxhall make them. The same goes for body type, 'Maestro hatchback' would be pointless as that is the only body derivative made. The next distinction is the engine size, usually expressed in cubic capacity, 1.3, 1.6 etc. the word 'litre' is another wasted one. Don't forget to include 'Turbo' or 'diesel' if that isn't part of the model description. The transmission is usually irrelevant, although an automatic gearbox is still unusual enough to merit a mention (many buyers have a preference for it) unless the car in question is a luxury model, where an 'auto' would be the norm. Most cars have five speed gearboxes these days, but it may be worth mentioning if fitted as an option, especially to smaller cars.

Many advertisers make the mistake of listing all the extras, but if you have identified the model, most buyers know what it comes with, so don't waste words unless your car is especially blessed with electric goodies. Everyone expects a radio, but perhaps not a CD. Equally the phrase 'all the usual extras' serves no useful purpose.

5. Colour. A crucial buying factor to many people. If metallic, this is worth mentioning; manufacturers names though are not always ideal, as they don't describe colour that well. When selling a banger this could be sacrificed if you are desperate to cut down the word rate.

6. Condition: Honesty really is the best policy. However there is no point in highlighting every blemish, people expect the odd scratch on a used car If you're too hard on your own car then no one will call. But if there are major faults, it is worth your while pointing them out so that buyers know what they are in for and you won't have streams of people after as new cars. There is no reason why a dodgy gearbox should make your car unsalable. 'Excellent condition, but gearbox needs attention hence that lovely word 'hence' qualifies the defect and justifies a lower than normal price. 'Suit enthusiast' is another way of describing a hopeless case. There are plenty of people out there who simply love getting their hands dirty, appeal to their instincts. So 'needs work for MOT' is another good phrase. Even if the car looks dreadful but is in perfect working order then 'cheap runaround' or 'ideal third car' work well.

7. Price: Sounds obvious, but plenty of advertisers put the word 'offers' or 'ono' (or near offer) at the end of the description. This is wrong. Firstly it indicates that you are desperate to sell at any price and two, however reasonable the asking price, someone will always want to make their own offer anyway. POA (Price on application) may add some mystique to the sale especially if the price when applied for is a steep one, but it won't deter timewasters and does not supply the crucial information that every buyer needs; namely can I afford it?

8. Telephone Number. How else is someone going to let you know they are interested by carrier pigeon? The most important thing is to make sure that you are there when the phone rings, otherwise buyers will give up. Clairvoyancy skills are not required, but a phrase like ring between 6-7pm means that you don't become a prisoner to the phone. Also, don't forget to include the STD code for your area, to make it easy for the buyer as they might go on to the next ad which doesn't require an all-night vigil waiting for directory enquiries to respond.

9. Owners: If you have been the only custodian, or there have only been a couple in the last decade, this is another reassuring indicator for the potential purchaser

10. Warranties: Any existing guarantees, or warranties, especially if provided by the manufacturer should be mentioned.

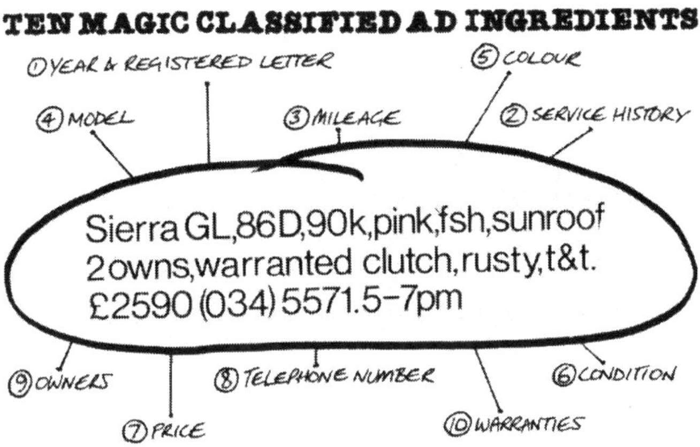

DEALING WITH BUYERS: Like it or not, the grumpy so and so kicking the tyres on your pride and joy is a customer Treat them as such and you stand a better chance of selling the car, but always stay in charge of the situation.

TELEPHONE: This will be your first contact with the potential buyer Be positive, words like average, fair and OK don't inspire confidence. Don't go silent on them and give as much information as possible. Don't negotiate the price over the phone.

THE APPOINTMENT. Make a precise time for the appointment and don't allow yourself to be trapped into an inconvenient time, or become a prisoner in the house waiting for someone to put in an appearance. Always take the enquirer's telephone number Never make the appointment late in the evening.

BUYER ARRIVES: Be friendly and helpful, remember you are not desperate to sell this car so don't fuss around the buyer, or talk too much.

THE CAR: Simply ensure that all the doors are unlocked, raise the bonnet and leave the buyer to it, but don't leave the keys with them! Then retire gracefully to prod at the garden, stay in the vicinity, but let the buyer look around. Don't offer to do anything, like start the car or go for a drive, until they ask. If they are interested then they will ask, if not they may go along with your offers to keep you happy. Also, leaving them with the car makes it less likely that they will criticise and put you on the defensive.

TEST DRIVE: The most important thing to remember is

that you should remain in control at all times and this can be achieved by sticking closely to the following guidelines:

INSURANCE COVER Often your insurance can cover other drivers provided you give your permission, but do you honestly want to jeopardise your no claims? Make it clear that if someone wants to drive, they have to be comprehensively covered for your vehicle and you will have to see a valid insurance policy as proof of this fact.

SECURITY One of the most worrying aspects of this whole test drive business is allowing someone who you would normally set the dogs on if they sauntered up the path, access to your car keys. Ask to see their driving licence and note down all the details (cross reference with the insurance policy), that will deter those with dishonest intentions and also make them treat your car with respect on the drive.

ALTERNATIVE DRIVE: If someone doesn't have the right insurance cover, or you just don't fancy them piloting your vehicle, the alternative is for you to drive them.

THE DRIVE:

1 Make sure that you drive first, even if the potential buyer wants a go and is so insured. This is so you can show that the car can be driven smoothly and there are no problems with a crunchy gearbox, juddering brakes and sluggish performance.

2. Drive to a quiet road invite the potential buyer to take the wheel. But before they do that, take the keys

with you, or they could take the car Make sure that they are comfortable, that the seat is adjusted properly and then talk them through the relevant controls, horn, lights, etc. Also set out the rules for the drive, say that you understand that they have to make sure that the car is in full working order but that you won't tolerate any arduous 'testing' When you are sure that they know where everything is, give them the keys.

3. Specify a route that leads you back home within a reasonable time. Chat freely during the drive, point things out that you think are relevant, but don't babble on. Silence often provokes favourable comments from the driver especially if the car is good and they begin the process of selling to themselves. Don't make any observations on the condition of the car, which could later be interpreted as a misrepresentation. Selling a car in an unroadworthy condition is an offence and leaves you liable for prosecution whether you knew about it or not. If you are aware of the defect, then you must make the shortcomings known to the buyer preferably in writing.

AFTER THE DRIVE: Ask if the buyer has any questions

and what they think about the car. This is also a good time to ask if they are going to buy. They have looked, touched, prodded and driven, so surely they want to buy? If not, why not. The objection may be a simple one that you can overcome. Perhaps they are worried about a noise, so offer to take it to garage with them to identify the trouble. Saying `If I can (...prove that the noise is normal and nothing to worry about...) will you buy the car?' may assist. Then if they finally say yes, your problems may have only just begun.

THE DEAL Be polite, low-key, helpful, never get excited and never show that you are desperate. Being chatty ought to help when it comes to the negotiations, which ought to be a pleasant exchange, and not a table thumping exercise. Bear in mind that you are not doing this professionally, there is no need for you to be slick and sophisticated. Be credible by accepting that your car isn't perfect in some small respect, then the buyer can't accuse you of being unreasonable.

GIVE & TAKE: Be prepared to drop in price to half of the difference, but don't do it until you're asked, never give away something for nothing. Don't let the buyer's silence pressure you into making a further reduction. Certainly never go below your limit, make a 50% reduction on that cushion and then down to the level.

SILENCE IS GOLDEN: Don't go over the top talking about the car, there is a tendency at this time to justify the price or the condition of the car Don't do it. Babbling weakens your position, gives too much away and shows that you are desperate to sell. Let the other person do that.

GET LOST. If someone does not want to pay your price, tell them politely that that is the deal, shake hands and say goodbye. Being polite always leaves the door open so that the buyer can come back. Never get heated or upset, whatever the provocation. Ask them to ring in 10 days time, if the car has not sold in that time, then perhaps something could be worked out.

YES, BUT .... To tip someone over the edge, ally your acceptance of an offer to a condition. OK I'll take that, but you have to take the car by Friday.

WIMPS: If you think that you are a weakling when it comes to negotiation then why not nominate a friend, or hard-nosed spouse to do the difficult bit? After showing the buyer around the car and doing the test drive, then 'hand over' the buyer with a 'I've got no head for figures, my friend/spouse/dog handles all that' This is a technique used by showrooms which softens up customers with a nice salesman and then squeezes out the cash with a nasty one.

SOLD!: Once you have shaken hands, the hard part is getting the money. You should also put the agreement into formal terms, even though an oral contract exists once you have agreed the price.

DEPOSIT. Insist on cash; it concentrates the buyer's mind on your car a cheque can be cancelled and a promise of payment isn't always fulfilled. A customer with second thoughts, or suddenly finding another car that takes their fancy, can soon fail to come up with the readies. If there is no deposit, there is no intention to purchase and no good faith. Specify that it is non-refundable. Don't believe someone who says that don't have a bean on them, as no one goes to buy a car without money. At the most, give them half an hour to find a cashpoint machine.

RECEIPT. To acknowledge the money you should give the buyer a receipt and whilst you are at it, enshrine the details of the deal on paper. Keep it simple, write down the facts about the car (registration, colour, make), the parties, (you and the buyer), the TERMS: The important part is for the car to be 'sold as seen' or 'sold as seen tried and approved' as you are not guaranteeing that It will be fault free for two years or anything stupid like that. Should the car be sold for spares, again make this clear as you can be prosecuted for selling an unroadworthy car. Any conditions that you have agreed should also be included. So if you've put the onus on the buyer to take the car the next day and pay in full, put this in the contract. If you've agreed to have the car tuned, again be specific otherwise the buyer could ask for a full service, so if you are just going to change the plugs and points, say so. Then both parties sign. Make sure you have a copy.

DON'T This seems obvious, but don't hand over any documentation, or the car until you have received full payment. Some unsavoury characters have asked to borrow registration documents so that they can tax the car when in fact they want to use it to sell a stolen vehicle of the same make. Don't keep all the documents in the glove compartment, or show the buyer where you keep them. Never let someone take the car away 'to show their mum' even though they don't have the registration documents it won't stop them selling it elsewhere, or nicking all the best bits.

PERSEVERE: Don't stop taking enquiries, the first buyer could fall through. So if people are still responding to your brilliantly worded ad, explain the position and ask to take their number as you will give them a call if the deal falls through. Never under any circumstances accept more than one deposit as you will get yourself into legal hot water.

CASH: The best way to get paid, especially if it is a relatively small amount. Always count the amount, as buyers make mistakes, not always innocently, and get someone else to double-check it, if the denominations are small.

BANKER'S DRAFT. Effectively this is the bank's own cheque and is as good as cash and is the only other fool proof way of accepting payment on the day of delivery. It will only cost the buyer a few pounds and a day or so to arrange.

BUILDING SOCIETY CHEQUE: A building society equivalent of a draft, but caution is required as stolen cheques are in circulation and it may be difficult to differentiate from a normal cheque.

CHEQUE: Never take a personal cheque, unless you clear it first (usually three to five working days), or pay for special clearance.

12. A TO B OF BANGERS: THE BEST BANGERS MONEY CAN BUY?

You have read an advertisement in the local paper but, although tempted by the attractive low price, you have no idea what it is and what the catch is. To help you out, here is the A-B of bangers aims to cover most, but certainly not all, cars that have been built or imported into the UK over the last 20 years or so. Some, at the time and of writing may not even be bangers, so likely candidates have been tipped. Once again, Good Luck.

1. ALFA ROMEO Throughout the 1960s,1970s and early 1980s, the Italian motor Industry gained a gained a reputation for building cars that disintegrated. The British climate and salted roads in winter hastened the demise of Alfa Romeos (sf Lancia and Fiat), although they would also come to a halt much sooner due to poor quality electrical and mechanical systems. On the credit side the sporting character of the cars still makes them exciting purchases. Their depreciation rate is alarming and this makes even recent models temptingly cheap. Even if you opted for a small saloon, spares and insurance costs are likely to be prohibitive. ALFASUD (1973-1984) Brilliantly exciting sporty hatch, but let down by shoddy construction and attacked unmercilessly by rust. SPRINT (1983-89) Good looking coupe with rust problems. ARNA (1984-86) Shortlived, (thankfully) concoction of Alfa engine inside Nissan Cherry body shell. ALFA 33 (1983-onwards) five-door hatchback, which has proved more durable and practical than other models. ALFETTA (1973-87) Sporty, but otherwise unremarkable, saloons, but coupe GTV now an appreciating, though often rusty, classic. GIULIETTA (1978-86) Based on Alfetta and inherited its rust. ALFA 75 (1986-onwards) Much improved Giulietta, better built, cheap but expensive to run. ALFA 6 & 90 (1980-86) Large and rather useless, few sold, but don't even take one for free.

2. AUDI Falling some way behind BMW and Mercedes in the Teutonic image stakes has helped the Audi to depreciate faster and reach the banger market much more quickly. However, these are quality cars, which are strong and reliable. These are penalised by higher insurance and parts costs, though many items are common to VWs too. 80 (1973-86) Reliable saloon, go for latest model you can afford. 90 (1984-86) Well built, bigger engined 80, few sold, but good value. 100 (1977-onwards) Spacious and comfortable, Avant five-door hatchback and estate even more so. 200 (1980-onwards) More luxurious 100, with Turbo. Cheap, but expensive to run. COUPE (1981-onwards) Four-seat comfort, but Quattro expensive to maintain and a burden to buy.

3. AUSTIN (BL, ROVER GROUP ETC) British cars lost their identity throughout the 1960s and 1970s, so identical cars were badged purely for marketing purposes. Some of the models mentioned here are now likely to be badged as Rovers. Otherwise a good marque for providing banger fodder, even though the majority of their cars are front-wheel drive since the 1970s. Parts are no problem and there are plenty of expired examples to pilfer from. A favourite with senior citizens who usually look after their cars in a fanatical and fastidious fashion who make ideal sellers. Insurance very cheap. MINI (1959-onwards) The ultimate town car, constant demand keeps prices high. Rust kills them, otherwise engines go on forever Absurdly cheap to run, loads of miles to the gallon and excellent parts availability. METRO (1980-onwards) Early cars suffer from rust, but cheaper than Mini for equivalent year and more room. ALLEGRO (1973-82) Ugly saloon and even uglier estate. Luckily, familiar mechanics mean cheap to run and easy to work on. MAESTRO (1983-onwards) Useful hatchback, but older cars need to be checked carefully, especially the MG model. MONTEGO (1984-onwards) Competent saloons and roomy estates, but like Maestro needs to be chosen carefully. MAXI (1969-81) Huge interior room though modestly proportioned hatchback. Buy one before the rot has set in. PRINCESS (1975-81) Lot of car for not much money. Wedge shape looks odd, early cars disasters, later hatchbacked AMBASSADOR (1982-84) more practical.

4. BMW Might seem odd to class them as bangers, yet the late 1970s and early 1980s models are just that. A cheap purchase price will be offset by the effort and expense required to keep it mobile. Insurance is obviously above average. 3 SERIES (1975-onwards) Originally a small saloon that grew into a four-door in 1984. Can be cheap, but ensure they are not about to expire. 5 SERIES (1975-onwards) As for 3 Series but much less style and even cheaper prices. 7 SERIES (1977-onwards) Very comfortable and sophisticated saloon, running costs horrendous, but could be better bet than an old Jag.

5. CITROEN To the British car buying public this name has always stood for 'weird, so give a wide berth' This is due to the complex nature of the suspension systems and quirkiness of the interior. Consequently the marque is an acquired taste. 2CV (year dot-1990) Huge cult following, but not the best reason to buy it. Prices for good cars stay high, essential to drive before you buy. VISA (1979-87) Cheap little hatchback, often overlooked, beware rust, diesel engine a bonus. LNA II (1983-85) Really a Peugeot 104, ideal town car. AX (1987-onwards) Flimsy but successful small hatchback.

GS/GSA (1970-84) Unique saloon, not for the faint hearted. Reliable until neglected when suspension plays up and engines fail. BX (1983-onwards) First really successful Citroen in the UK, not brilliantly built and still with 'that' suspension. Check carefully. CX (1975-89) Ultimate weird Citroen, huge amount of space (especially estates) full of extras, excellent ride, cost nothing to buy, but a nightmare when they go wrong.

6. DACIA What has the Rumanian Motor industry ever given the world? Answer: the Dacia. Quality not good when new, so could be appalling when a banger DENEM (1982-83) Really a Renault 12. Must be cheap to make sense. DUSTER (1984-onwards) Four-wheel drive, Renault powered utility type vehicles. Again, must be cheap.

7. DAIHATSU Well built cars, but never marketed extensively in the UK, so getting one fixed could be difficult and the parts expensive. DOMINO (1981-onwards) Tiny hatchback, with tiny engine, suitable only for tots. New model from 1986 much better. CHARADE (1980-onwards) Reasonable, but noisy and small five-door hatchback. New 1987 models slight improvement. CHARMANT (1982-87) Anonymous saloon. Few sold.

8.DAIMLER See **Jaguar**. Note: same car, different badges, and few more bits of chrome. Good news is that prices are often less than its relative because Flash Harry would rather have a Jag!

9. FSO/POLSKI Rare in this country and arguably a poorer proposition than a Lada. Cars based on redundant Fiats, so there is not much to recommend them, as they are slow, unrefined and parts expensive.

For short-term use only, then they must be thrown away. 125P (1975-onwards) A 125 Fiat saloon built in Poland, tough but crude. POLONEZ (1979-onwards) Slightly more successful model, basically a large five-door hatchback, must be dirt cheap to be good value.

10. FIAT Although Fiat suffered the same rust slurs as the other Italian manufacturers their cars made more successful inroads into the mass market, especially at the smaller end of the scale. Even so these cars are very flimsy and likely to be held together by string at the banger stage of their lives. 126 (1973-onwards) Smaller than a Mini but with none of its charm or performance. Rust is a big enemy. Polish built Bis from 1987 had hatchback and was a slight improvement over the original. 127 (1971-83) Originally a saloon that turned into a hatch, rust has reduced choice. 128 (1969-82) Small saloon and estates with enthusiastic following. Getting rare. PANDA (1981-onwards) Very basic town car, rust prone and unrefined but does a million miles to the gallon. UNO (1983-onwards) Best Fiat for years, can be noisy and unreliable if neglected. Versatile hatchback best in class. TIPO (1988-onwards) Another good Fiat, roomy and modern, prices stay high. TEMPERA (1990-onwards) Characterless saloon, getting cheaper. STRADA (197988) Hatchback that rusts, but at least it is cheap. REGATA (1984-90) Saloon version of Strada, ignored in UK, so very cheap. X1/9 (1977-89) Impractical coupe but fun and still very cheap. This baby Ferrari rusts though. FIAT 131/MIAFIORI (1975-84) More dull saloons and estates, poorly made but at giveaway prices. FIAT 132/ARGENTA as for 131 but larger engines. CROMA (1986-onwards) Attempt at an executive model, but few takers, so very cheap, build quality poor

11. FORD The quintessential banger marque, Ford stands for cheapness, reliability and availability. In theory one of the cheapest of all cars to run. However, their popularity results in higher prices than the average banger so by the time they reach rock bottom, the cars are often worn out. Bear this in mind when choosing. FIESTA (1977-onwards) Not surprisingly the best-selling small hatchback. 1983 re-styled models are probably the best. ESCORT (1968-onwards) Has always been with us, Mark 2 from 1975 a good place to start for basic motoring requirements. Mark 3 manufactured from 198C had front wheel drive, but no less reliable or practical. ORION from 1983 booted four-door version of the hatchback. CORTINA (197082) The original Dagenham Dustbin dates back to 1962, but there are few survivors. As bangers go, this has it all, with plenty of room and vroom from the bigger engined versions. Don't buy anything smaller than a 1.6, whilst Ghias promise luxury specification Watch for rust and don't pay through the nose for one of the best bangers about. CAPRI (1969-86) Essentially a Cortina with knob on, but less doors and less room. 1978 hatchback modification useful for some and bigger engined cars more exciting. SIERRA (1982 onwards) Cortina replacement and, although not everyone liked the jelly mould shape, it is a worthy successor Same rear drive layout and engines mean cheap, maintenance and parts. Hatchback joined by saloon version called SAPPHIRE in 1987 GRANADA (1972-onwards) Armchair comfort and big fuel bills, but utterly reliable with good levels of equipment. Mark 2, 1977-1985, is the best' banger, with good estates, and is most sought after. The latest less popular five-door hatchback shape are now becoming affordable.

12. HONDA 'Made in Japan' was originally a derisory term, implying cheap and poor quality. Their cars though proved to be relentlessly reliable even if they did rust very badly. Honda may now be in a different class, but their older models seem to go on forever the only drawbacks are those associated with foreign ownership, namely insurance and parts. JAZZ (1984-85) Small town car Good mpg. CIVIC (1973-onwards) Ever evolving range of cars, but reliable small hatchbacks and estates are well built. BALLADE (1987-89) Sturdy small saloon also badged as a Rover 213/216. INTEGRA (1986-89) Five-door hatchback that proved reliable and popular ACCORD (1978-89) Saloons that grew up over the years. Initially rust prone, the post-1985 cars have more room and AERODECK a coupe/estate that is highly desirable. PRELUDE (1979-onwards) Story for this coupe is as Accord on which it was based. However, a Bangernomics student might question whether four-wheel steering is necessary, or cheap to repair LEGEND

(1986-onwards) Japanese version of the Rover 800, which never caught on and so prices are low.

13. JAGUAR For a long time Jaguar saloons have followed the same pattern by depreciating rapidly and becoming a lot of flash for not much cash. Of course there is likely to be something wrong with it at a banger price, as keeping any Jag/Daimler running is horrendously expensive. However if you don't do many miles the old girl may cling to life for longer than you thought. Be prepared to throw away once it breaks down. XJ6 (1968-86) Comfortable saloon, with silken smooth power is the theory. In practice you can end up with a pile of expensive junk. There is a lot to go wrong if it has not been looked after. But strike it lucky and you get an awful lot of saloon for your money. XJ12/DAIMLER DOUBLE SIX (1972-onwards) Not only have you got two fuel tanks and exhaust systems to worry about, but also two lots of 6 cylinders, so there is even more that could go wrong. Fuel consumption is horrendous, but what a way to go bankrupt. XJS (1975-onwards) This coupe also seems too cheap to be true and then you get knocked for six with the first repair bill.

14. LADA Cheap and nasty Eastern Bloc bangers. It does not matter whether they are new or used. Secondhand they look like giveaways if you can live with their shortcomings. Often they harbour niggly electrical problems and more rattles than a nursery due to poor quality construction. Otherwise tough, though parts can be difficult to find and costly. The newest banger you can buy. 1200/1300/1500/1600/ RIVA (1974-onwards) A Fiat 124, but slightly more durable. Watch out for rust. SAMARA (1987-onwards) New generation front wheel drive car, streets ahead of RIVA, but still crude. NIVA (1978-onwards) Russian Range Rover tough but basic.

15. LANCIA Like Alfa Romeo, but much worse. Lancia's talent for disintegration due to rust was legendary. Also like Alfa an oddball mixture of sporty engines and oddball interiors. Y10 (1985-onwards) Good attempt at a toy town hatch, but unrefined and little room for more than two passengers. Reassuringly cheap. DELTA (1980-onwards) Five-door hatchback that has won rallies, but in the real world can be unreliable. PRISMA (1983-90) saloon version of Delta. BETA (1976-85) Saloons and coupes that rusted dangerously in the early years. THEMA (1985) underrated executive saloon, which can be full of problems.

16. LAND ROVER If you really must stray from the Queen's Highway then these are probably the first cars you consider doing it in. Land Rovers rarely get scrapped and seem to live forever, because every single part is replaceable. That means you get what you pay for. There is no such thing as a really cheap Land Rover. LAND ROVER (1948-onwards) There have been several versions of the lovable 'Landie' they are all slow, thirsty and uncomfortable, so make sure you really need the four-wheel drive. Don't buy ex-farmyard or building site hacks. RANGE ROVER (1970-onwards) Up-market estate car does zero miles to the gallon and can be expensive to put right. They rust too. Still want one?

17. MAZDA As Honda, but has a much smaller presence in this country, consequently priced to please Bangernomic students. 323 (1979-onwards) Front-wheel drive hatchbacks and estates, slightly unrefined for some, huge choice of models. Early cars rust. 121 (1988-onwards) Reliable small hatchback. MONTROSE (1979-81) Shortlived Cortina competitor 626 (1981-onwards) A large range of reliable saloons, hatchbacks and coupes.

18. MERCEDES The first, and probably last, word in German efficiency. These are no frills cars that happily work in the Fatherland as Taxis yet are perceived everywhere else as upmarket executive expresses. They last forever, so depreciate slowly, but ultimately live long enough to become bangers. 200-300 (1976-86) Tough saloons and estates. So-called baby Merc, holds prices well, but now more affordable. S-history is essential, especially with the new generation (1986-CLASS (1973-onwards) Huge saloons which are virtually bullet-onwards) models. Diesels are highly prized. 190 (1983-onwards) proof, but cost a King's ransom to run.

19. MITSUBISHI Initially known as Colt in the UK market. As Honda and Mazda with excellent reputation for producing reliable cars. Again, only limited numbers imported. TREDIA (1982-85) Saloon range which never caught on, so rare and cheap. CORDIA (1982-88) Coupe, which shared same fate as the Tredia. LANCER (1984-onwards) Saloons and estates which got better as time went on; from 1988 hatchbacks and sporting versions too. COLT (1984-onwards) Small hatchback range, which got larger in 1988 and shared reliable Lancer engines. GALANT (1974-onwards) Around for ages but only take seriously from 1980, this medium/large saloon is well equipped and utterly reliable. SIGMA (1983-85) Australian-built saloons and estates briefly sold as the Lonsdale. SAPPORO (1981-89) Originally a brand name for a coupe that was discontinued in 1985, then from 1987-89 as upmarket Galant saloon. SPACE WAGON (1984-onwards) Competent 'people carrier' prices stay high for this van with windows. SHOGUN (1983-onwards) An early entrant into the small four wheel drive off-road market, spacious and reliable.

20. MORRIS In many ways as Austin. Using the Morris name was part of BL's attempt to give their models more character MINOR (1948-1971) Ancient saloon worth mentioning because so many are still about. You can pay a fortune, but often there is no need, especially for the less trendy four-door saloons. Easy DIY and excellent parts availability makes this a banger with real style and if you look after it, the Minor won't depreciate. MARINA/ITAL (1971-83) Has many of the virtues of the Minor, easy DIY cheap parts and also the vices, such as rust. Keeps up with traffic better than Minor and is more comfortable, not everyone's cup of tea, but otherwise competent saloons and estates.

21. NISSAN/DATSUN Originally marketed in the UK as Datsun. See Honda, with even wider availability and models to suit all tastes. Like many of the Japanese manufacturers, Nissan has gone up market, but their older cars provide a remarkable amount of potential for Bangernomics purposes. However, most mid-1970s models have now rusted to dust, although the engines are still likely to be running. Conclusion: Alternative Oriental Ford. CHERRY (1982-86) Not the best Nissan, as hatchback very rust prone. SUNNY (1982-onwards) Large range of cars from saloons and estates to coupes, all comfortable and reliable, though the older ones look their age. MICRA (1983-onwards) One of the best small hatchbacks, thousands of driving schools can't be wrong, just don't buy from them. STANZA (1982-86) Hatchbacks and saloons, which are competent and reliable. BLUEBIRD (1980-onwards) Saloons and estates that are genuine Cortina/Sierra alternatives. PRAIRIE (1983-onwards) People carrier with plenty of room. SILVIA (1984-89) Coupe; watch those insurance premiums. LAUREL (1979-89) Large saloon which rusted in the early years, but had high specification. MAXIMA (1989-onwards) took over from where Laurel left off, but not popular, so lots of car for not much money. PATROL (1982-onwards) Four wheel drive is happier when off road. Diesels at a premium.

22. OPEL These were up-market Vauxhalls, built in Germany, but recently only a few cars were badged exclusively as Opels. MANTA (1980-88) The only real rival to the Capri, either as hatchback, or coupe. Check that it has not been smashed and cobbled back together before you buy. SENATOR (1978-84) Luxurious rival to Granada. Well equipped, it turned into a Vauxhall in 1984. Beware ex-taxis. MONZA (1978-87)

Coupe version of Senator, which gulped fuel and rivalled BMW for top coupe slot. Now at giveaway prices.

23. PEUGEOT A manufacturer whose range, with notable exceptions, seem to depreciate rapidly whether the car is good or not. This is mostly due to poor build quality. Best known for their larger cars, diesels and estates which hold their values best. 104 (1980-83) Small, almost pointless, hatchback. Cheap and rust prone. 205 (1983-onwards) Peugeot show the world how a small car should be. Economical and reliable, if flimsy. Prices stay firm, especially diesels. 305 (1978-88) Competent and cheap medium sized saloons and hatchbacks. 309 (1986-on) Hatchback built in UK, but with chequered reliability history. 405 (1988-onwards) Another Brit-built Pug, more exciting than a Sierra, but once again not that solid a saloon. 504 (1968-82) Big saloon and estate. Rust loves them. 505 (1979-91) Another big success story, especially the diesels and huge estates. Saloons are the best bargains. 604 (1975-83) French idea of luxury, bags of equipment, but worth nothing, so a great bargain barge.

24. PORSCHE Having fallen out of yuppie favour, many people now see the cars for what they really are. The problem is insurance and running costs. 924 (1977-88) the first affordable Porsche, largely made up of VW and Audi parts. Some examples very cheap, but watch out for accident damaged ones.

25. PROTON By going expensively upmarket the Japanese left behind a gap for the Malaysians to buy the old Mitsubishi Lancer (sf above) and make the PROTON (1989-onwards) to the same quality at a lower price.

These are reliable cars but can be hard to find.

26. RELIANT Best known for the infamous three-wheelers. The plastic body means no rust. Fortunately the cars are difficult to find. RIALTO: The three-wheeler which is noisy, has uncertain handling, but as a bonus uses little fuel. KITTEN (1976-1982) Plastic body, but this time a wheel at each corner and mini engine. Almost recommended. SCIMITAR GTE/GTC: Reliant's saving grace, the Ford-powered coupes and convertibles which can be temperamental.

27. RENAULT As Peugeot in many ways. Rust and flimsy construction don't help matters when they reach the banger stage. Basic examples offer more charm and chance of reliability. 4 (1962-86) Renault's answer to the 2CV more popular in its native land too, for this rust prone but attractively basic runabout. 6 (1968-79) as 4 but more powerful engine and more room, few about. 5 (1972-onwards) Rust is the major enemy, otherwise this trend setting hatch is the best baby banger you can buy. 9 (1982-89) small saloon, cheap, dull and not always reliable. 11 (1983-89) Hatchback version of the 9, so more useful, but not more interesting, except Turbo which needs to be checked thoroughly. 18 (1979-86) Saloon and estate cars that are dull, but competent. FUEGO (1980-86) A Renault that was not a number, but that did not make it any good. At least you can get four people inside this coupe. Otherwise has to be cheap to be good value. 20 & 30 (1975-84) Big spacious and often well-equipped hatchbacks, but rusty and unreliable. 21 (1986-onwards) Saloons hatchbacks and estates, which are competent and comfortable. 25 (1984-on) Big, comfortable, well-equipped hatchback that depreciates quickly. ESPACE (1985-onwards) Best people carrier for many years, has rot resistant glass fibre body and is better built than most of the other models, so prices stay high.

28. ROVER Although all former British Leyland products are now badged as Rovers, the marque also had an identity of its own prior to this. SD1 (1976-87) Large executive hatchback, which looked good but had plenty of teething troubles. Thirsty V8 is the most reliable buy whilst the six-cylinder 2300-2600 is troublesome. Early ones are now very shabby. 213 & 216 (1984-90) Rover version of the Honda Ballade, these small saloons are well finished and lively. 800 (1986-onwards) Large saloon and later hatchback to replace SD1 They depreciate heavily and offer comfort with high equipment levels. 214/216 & 414/416 (1989-onwards) Rover's better version of the Honda Concerto comes as a hatchback or saloon.

29. SAAB Regarded by some as the Swedish BMW there is no denying that these are tough cars with character. Expensive to put right though and old cars with lots of extras can play up. Don't get mixed up with a Turbo-charged cheapie. 99 (1970-84) Solid saloons and hatch-backs. Good reputation for reliability, but watch the prices of parts and servicing. 900 (1979) Hatchback with limited interior space, only buy Turbo that has been looked after. 9000/CD (1985-onwards) Large executive hatchback and saloon, which depreciates quickly.

30. SEAT European motor industry collides and co-operates together in Spain to build cars that are cheap, cheerful and worth next to nothing on the used market, so a good buy. MARBELLA (1988-onwards) A Fiat Panda, but not as good as the real thing. IBIZA (1985-on) hatchback that suffers from having no image and shaky build quality. A good banger. MALAGA (1985-onwards) saloon Ibiza, even better banger prospect.

31. SKODA You have heard all the jokes, but Skodas are not to be dismissed that lightly, being the best of the old Eastern Bloc mobiles. Obviously cheap when new, they are also giveaways when used and are probably the most reliable and durable Easterners you can buy. ESTELLE & RAPID (1977-90) Rear engined saloon and coupe. Go for 1980s cars which had improved and much safer handling. FAVORIT (1989-onwards) The Skoda for the 1990s, still unrefined but a nice looking hatchback body and front wheel drive.

32. SUBARU Another of the small Japanese companies who have made an impact in certain sectors of the market. 1600-1800 (1980-90) Range of saloons and estates with four wheel drive, popular with farmers as durable and spacious, but don't buy their cast offs. JUSTY (1986) Small four-wheel drive hatch, interesting, but do you need 4WD? XT TURBO COUPE (1985-89) An oddity that flopped, so prices are on the floor, don't be tempted by this ugly duckling.

33. SUZUKI Yet another tiny Japanese concern building sound, reliable cars, that is difficult to fault. ALTO (1979-86) Small saloon and hatchback, but unrefined. Parts could be a problem. SWIFT (1984-onwards) range of competent hatchbacks that hold their value. SJ & SAMURAI (1982-onwards) Four wheel drive 'fun' Jeepish vehicle. Neither one thing, nor the other clever marketing kept prices high, but was reported as dangerous if used in anger. Pay little and use carefully.

34. TALBOT/CHRYSLER Confusion of model names, badges and owners throughout the 1970s when a lot of old Rootes Group companies amalgamated to become Chrysler, then taken over by Peugeot and renamed Talbot. Lets try and unravel the mess to reveal some of the great bangers. ALPINE (1976-85) Large hatchback that rusted badly, but had a tough engine. Brilliantly cheap. HORIZON (1978-85) medium sized hatched, badged as a Simca in Europe, with a strong engine and liking for rust. SAMBA (1982-86) Yet another version of the Peugeot 104. Change of badge does not improve reliability or rust resistance, but cheap and cheerful. SOLARA (1980-85) Booted Alpine with all the same attractive features. TAGORA (1981-84) Attempt to build a large executive saloon, a total disaster. The few that were sold are now amongst the best equipped bangers you can buy. RANCHO (1978-85) Endearing crack at the Range Rover market, but without four-wheel drive. Rear body made of rot free glassfibre, but the rest isn't.

Engine tough, and roomy interior makes for a great banger estate. AVENGER (1970-81) Classic Banger layout of tough engine and rear wheel drive. Can become flaky at the edges, but has survived well. SUNBEAM (1977-82) Reworked Avenger, now smaller with a hatch, but still has rust. MINX/HUNTER (1967-79) Like Avenger, but there seems to be fewer of this Cortina rival about. Rust has finished most off.

35. TOYOTA Toyota vies with Honda for the title 'BMW of the Orient' Well built, reliable cars that have become increasingly expensive. Great bangers because of their ability to go on forever, provided you don't pay too much. STARLET (1978-onwards) Small, reliable and economical hatchbacks. Earliest cars now rusting nicely, so they should be cheap. TERCEL (1982-88) More useful hatchbacks and even a four wheel drive estate which could be an unnecessary complication. COROLLA (1980-onwards) Huge range of hatch, saloon, estate and four-wheel drive estate, from economy to performance models. Go for the simpler earlier incarnations. CARINA (1978-onwards) More of the same, but slightly larger. CAMRY (1983-onwards) Good all-round large saloon and estates. CELICA & SUPRA (1980-onwards) Sports coupes that are reliable and expensive to run. The Supra is closest in spirit to the Capri, insurance likely to be frightening. LANDCRUISER/SPACECRUISER (1981-onwards) People carriers that in the case of the huge land Cruiser could accommodate a football squad. The good news is they are also rugged and reliable.

36. TRIUMPH Before BL died Triumph was allowed to make and sell its own cars, up to a point. TOLEDO/DOLOMITE (1969-81) Classic rear-wheel drive banger layout, but shoddy build quality let otherwise pleasingly quaint British package down. Shop carefully to avoid rot and unreliability, otherwise perfect. TR7 (1976-82) Worth mentioning as the only TR normal people can afford and a real sporting banger Just a Dolomite in drag really and not that good to look at, except convertibles. When they are cheap, they are cheap for a reason. Beware. ACCLAIM (1981-84) Rebadged Honda Accord, built from bits in Britain. Most reliable Triumph for a long time. Body rots, but well looked after examples are worth finding.

37. VAUXHALL The other Ford, except they are owned by General Motors, producing a huge range of tough, reliable, easily serviced and insured cars with its fair share of potential and actual bangers. VIVA (1966-79) A great survivor with a large number of saloon, estate and coupe offspring. Engine goes on forever whilst the body moulders with rust. Worth having. CHEVETTE (1975-82) Viva relative, either as saloon, or useful hatchback/estate. DIY a doddle but rust abundant. CAVALIER (1975-onwards) The other Cortina. In its original form had rear-wheel drive, but restyle and overhaul in 1981 meant front-drive from then on. Saloons hatchbacks and estate largely indestructible. ASTRA (1980-onwards) Competent all round, hatchback and estate workhorse. Total restyle in 1984 meant new slippery shape and so more to pay. Less popular but spacious Belmont saloons from 1986. NOVA (1983-onwards) Small hatch or saloon, reliable and popular, so prices stay high. CARLTON (1978-onwards) Big executive saloon and huge estate. Extra luxurious Viceroy, Royale and Senators with larger engines. Depreciation awesome, so stress free banger motoring provided not corroded, or an old taxi.

38. VOLKSWAGEN A bye-word for reliability has an almost unassailable reputation. However, neglected cars sold by the unscrupulous, bite back, also, watch parts and servicing costs, which can mount up. BEETLE (year dot) Worth a mention as it is the car most often used to illustrate VW's reputation for reliability. In truth, overrated, not always appropriate or comfortable for modern motoring conditions. Far too many sellers asking optimistic 'classic' prices. The right vehicle would make a good buy provided the dreaded rot has not struck. GOLF (1974-onwards) Best ever hatchback. Got bigger in 1984. Almost faultless though rust corrodes early cars and high miles, means smoky engines. Hard to find sometimes at the right price as satisfied owners keep them forever. JETTA from 1980 booted version, not as popular so cheaper but still plenty of room. POLO (1976-onwards) Small hatchback, saloon and coupe equally well built as bigger Golf. PASSAT (1981-onwards) Overlooked saloon, hatchback and estate, both, solid and reliable. Dirt-cheap. SCIROCCO (1975-onwards) stylish Golf-based coupe. Whilst most would rather have the more practical Golf GTi, Sciroccos are a steal, but watch the insurance group.

39. VOLVO You could be forgiven for thinking that they build tanks. But no, this Swedish company makes rather large, dull cars that have been marketed as being extremely safe. 240/244/245 (1974-onwards) Second generation tank with huge bumpers. The estates are always sought after especially by builders and antique dealers, so saloons can go begging. Reliable and comfortable if boring. 700 (1982) More modern looking tanks, estates even more desirable, saloons just a cheap and boring, but acres of space. 340/343/345/360 (1976-onwards) Because they are Volvos people think they are good, not true. Hatchback is not that roomy and generally it is nasty to drive. Not always cheap because of Volvo kudos, buy when no longer shiny and you'll get a bargain.

YUGO Also known as Zastava, this is another crude and underdeveloped botch-up from the East, this time Yugoslavia (or what is left of it). Only buy if very desperate. 45/55/65 (1983-onwards) warmed-over Fiat 127-128 but much more badly built. 1100/311/511 & 1300 (1983-onwards) This time they had a crack at the Fiat 128 with equally disastrous results. 411 & 413 were saloon versions, why bother? SANA (1990-on) almost acceptable, but still basic Fiat based hatchback.

38.
39.
40.

13. BANGLOSSARY: Checklists

Here they are, all the crib sheets, checklists and standard forms you will need to help you become fully qualified in the art of Bangernomics. The author and publishers grant permission to the owner of this book for copies to made for your own private purposes, not for resale, or other distribution.

BANGERNOMICS CHECKLIST: 1. PAPERWORK

V5 INFORMATION

MAKE:

MODEL.

COLOUR:

REGISTRATION NO:

ENGINE NO:

CHASSIS NO:

NUMBER OF PREVIOUS OWNERS:

PREVIOUS OWNER: NAME

ADDRESS

DATE OF REGISTRATION:

DATE OF LAST SALE:

BILL OF SALE/PURCHASE RECEIPT YES/NO

SERVICE HISTORY/BILLS YES/NO

DETAILS:

LAST DATE SERVICED:

NEXT SERVICE DUE:

MOT CERTIFICATE

DATE ISSUED:

MILEAGE:

REGISTRATION NO:

CHASSIS NO:

AUTHENTICATION STAMP.

ARE THERE ANY PREVIOUS CERTIFICATES:

DO THE DETAILS CORRESPOND WITH V5: YES/NO

ROAD FUND LICENCE VALID UNTIL

BANGERNOMICS CHECKLIST: 2. EXTERIOR/BODYWORK

DOES THE CAR SIT LEVEL ON LEVEL GROUND?

FRONT YES/NO

SIDE YES/NO

GENERAL APPEARANCE: EXCELLENT/GOOD/AVERAGE/POOR INDICATE ANY DAMAGE:

KEY. R = RUST; S = SCRATCH; R/S = RE-SPRAYED; D = DENT. C = CRACK

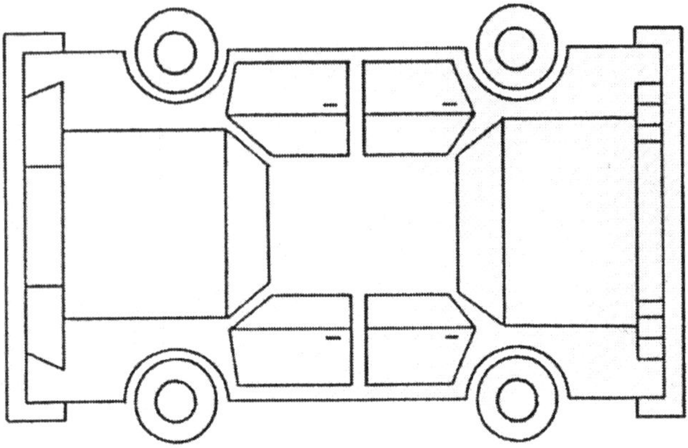

TYRES NEW GOOD POOR DAMAGED

FRONT LEFT

FRONT RIGHT

REAR LEFT

REAR RIGHT

SPARE

REAR LEFT

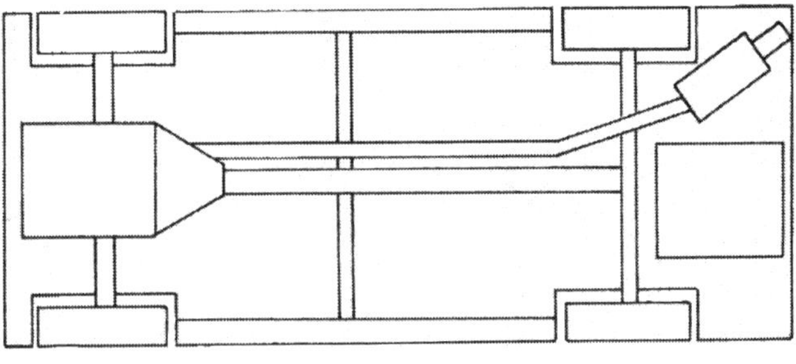

BANGERNOMICS CHECKLIST: 3. UNDERSIDE

INDICATE: L = LEAKS R = RUST D = DAMAGE OIL ON GROUND WHERE PARKED? COMMENTS:

SUSPENSION BOUNCE TEST.

FRONT LEFT YES/NO

FRONT RIGHT YES/NO

EAR RIGHT YES/NO

REAR LEFT YES/NO

BANGERNOMICS CHECKLIST: 4. UNDER BONNET ENGINE

ENGINE COLD:

GENERAL APPEARANCE:

OIL LEAKS:

ENGINE NUMBER: (Check against numbers taken from V5)

ENGINE OIL LEVEL. CORRECT YES/NO

CONDITION

AUTOMATIC GEARBOX OIL LEVEL. CORRECT YES/NO

CONDITION

VIN/CHASSIS NUMBER:

WATER LEVEL: CORRECT YES/NO

CONDITION

BATTERY LEVELS (Unless sealed): CORRECT YES/NO

CONDITION

R = RUST

ENGINE RUNNING

NOISES

EXHAUST. NOISES SMOKE

BANGERNOMICS CHECKLIST: 5. INTERIOR

OVERALL IMPRESSION GOOD/AVERAGE/POOR

CARPETS • ROOF LINING • SEATS • DOOR TRIM • DASHBOARD • SEATBELTS• MILEAGE • DOES CONDITION OF INTERIOR RELATE TO MILEAGE? SPECIFICATION: WHAT WORKS? • RADIO/CASSETTE • SUNROOF • HEATER • LIGHTS • INDICATORS • SIDELIGHTS • HEADLIGHTS • FOGLAMPS• STOPLIGHTS • WINDSCREEN WIPERS • WINDSCREEN WASHERS • AIR CONDITIONING • ELECTRIC WINDOWS
OTHER EXTRAS:

BOOT • SPARE WHEEL • TOOLKIT • WHEEL BRACE & JACK • RUST

BANGERNOMICS CHECKLIST: 6. TEST DRIVE YES/NO

STARTS EASILY? • OIL LIGHT GOES OUT • IGNITION LIGHT GOES OUT

PULL AWAY • CLONKING SOUND • JUMPS OUT OF GEAR • SLUGGISH PERFORMANCE • HILL. - JUMPS OUT OF

GEAR • HILL START • HANDBRAKE HOLDS • DESCENDING EXHAUST SMOKE • EMERGENCY STOP • PULL TO SIDE • SQUEAL • PULL AWAY IN SECOND: • STALL • KNOCKING

STEERING • WANDERS WHEN GRIP RELEASED ON ROUGH GROUND: • CAR PITCHES • BANGING NOISES CORNERING: • CLICKING • CLONKS/KNOCKS • RUMBLE FROM WHEELS

DUAL CARRIAGEWAY • VIBRATION • SMOOTH AUTOMATIC GEAR CHANGE

AFTER THE DRIVE: • OIL LEAKS IN ENGINE BAY (engine running) • LEAKS FROM WATER HOSES • EXCESSIVE SMOKE FROM OIL FILTER

7. BANGERNOMICS RECEIPT

I (SELLERS NAME)

of (ADDRESS)

confirm that I am the owner of the vehicle

MAKE MODEL

REGISTRATION (or that I am authorised to sell it on behalf of the owner NAME ADDRESS.

I confirm that the vehicle does not have any faults, which I have not disclosed/and that the recorded mileage of is correct.

I agree to: (DETAIL REPAIRS)

before sale.

8. HOLDING DEPOSIT

I accept the sum of £ as a holding deposit on the above vehicle, which is fully returnable provided a decision is made within 24 hours.

The purchase price for the above vehicle will be £

SIGNED

DATE

14. BANGER FACT FILE: USEFUL ADDRESSES

Here they are, a number of organisations that might help you side step some problems on the way to hassle free banger ownership.

HPI AUTODATA: Computer records with information on whether car is stolen, still subject to a finance agreement, or has been written off. A charge is made for this service. Contact. HPI Information, Dolphin House, PO Box 61 New Street, Salisbury, Wilts SP1 2TB.

RETAIL MOTOR INDUSTRY FEDERATION: Complaints about member garages. Contact Customer Relations, RMIF, 9 North Street, Rugby CV21 2AB. Telephone: 0788 576465.

SCOTTISH MOTOR TRADE ASSOCIATION: Complaints about member garages in Scotland. Contact SMTA, 3 Palmerston Place, Edinburgh EH12 5QA. Telephone: 031 225 3643

INSTITUTE OF THE MOTOR INDUSTRY. Complaints about individual members who are garage staff. Contact: Fanshaws, Bickendon, Hertford SG13 8PQ.

Telephone: 0992 86521

VEHICLE BUILDERS AND REPAIRERS ASSOCIATION: Complaints about bodywork repairs by members only. Contact VBRA, Belmont House, Gildersome, Leeds LS27 7TW. Telephone: 0532 538333.

WHICH? PERSONAL SERVICE: Members only will get legal help for a fee. Contact. Consumers' Association, 2 Marylebone Road, London NW1 4DX. Telephone: 071-935 3277

AUTOMOBILE ASSOCIATION: Vehicle inspection for members and non-members, cost varies accordingly and depends on complexity of car Contact 0345 500600.

Free legal advice for motoring problems, members only. Contact Legal Department, AA, Fanum House, Basingstoke, Hants RG21 2EA. Telephone: 0256 20123.

ROYAL AUTOMOBILE CLUB: Vehicle inspection for members and non-members, cost varies accordingly and depends on complexity of car Contact 0800 3336660.

Free legal advice on motoring matters for members only. Contact. RAC Motoring Services, RAC House, M1 Cross, Brent Terrace, London NW2 1LT Telephone: 081-452 8000.

SOCIETY OF MOTOR AUCTIONS: Information about members who abide by a code of practice in your area. Contact SMA, PO Box 13, Wilmslow PDO, Cheshire SK9 1 LL. Telephone: 0625 536937

15. DEGREE IN BANGERNOMICS; READ THE BOOK, BUY A BANGER AND GET SOME LETTERS AFTER YOUR NAME!

If you would like a genuine Bangernomics proficiency certificate, all you have to do is buy a car, then send proof of purchase (of Bangernomics the book) and a photograph of your banger to the address below (with a large stamped addressed envelope no certificate without it) and you will receive the certificate within 28 days. Be the envy of friends, family and people who hardly know you.

Before you embark on this great motoring adventure, remember these Bangernomics commandments:

1. NEVER SPEND MORE THAN YOUR BUDGET

2. DON'T PANIC BUY DO YOUR RESEARCH.

3. WHEN IN DOUBT GET EXPERT SECOND OPINION, OR FORGET IT

4. WALK AWAY IF YOU ARE NOT HAPPY, THERE ARE PLENTY MORE BANGERS TO CHOOSE FROM.

5. BEWARE OF THE DOG

ABOUT THE AUTHOR

Back in 1993 James Ruppert had written Dealing with Car Dealers, The Vauxhall, Jaguar and BMW Driver's Handbooks. A regular in Car Magazine and The Independent Newspaper he also operated under the pseudonym of Eddie Riff as an international man of mystery appearing in several publications. Including Performance Car.

Printed in Dunstable, United Kingdom

70668379R00113